Ferrets

Ferrets As Pets

Ferret Care, Keeping, Pros and diet and health.

By
George Galloway

Published by: Zoodoo Publishing

Table of Contents

Introduction 4

Chapter 2. Owning a ferret 24

Chapter 3. Decoding the ferret's behaviour 37

Chapter 4. Setting the ferret's home 46

Chapter 5. Ferret proofing the house 53

Chapter 6. Diet requirements of the ferret 59

Chapter 7. Health of the ferret 70

Chapter 8. Grooming the ferret 84

Chapter 9. Training the ferret 90

Conclusion 100

References 102

Introduction

I want to thank you and congratulate you for buying the book 'Ferrets as Pets'. This book will help you to understand everything you need to know about domesticating a ferret. You will learn all the aspects related to raising the ferret successfully at home. You will be able to understand the pros and cons, behaviour, basic care, keeping, housing, diet and health related to the animal. There are people who are impressed by the adorable looks of the ferret. They think that this reason is enough to domesticate the animal. But, domestication of a ferret has its unique challenges and issues.

If you are not ready for these challenges, then you are not ready to domesticate the animal. If you have already bought or adopted a ferret, even then you need to understand your pet so that you take care of him in a better way. It is important that you understand that owning any pet will have its advantages and disadvantages. You should see whether, with all its pros and cons, the animal fits well into your household. Domesticating and taming a pet is not only fun. There is a lot of hard work that goes into it.

It is important that you are ready to commit before you decide to domesticate the animal. If you are a prospective buyer, then understanding of these points will help you to make a wise decision. When you bring a pet home, it becomes your responsibility to raise the pet in the best way possible. You have to provide physically, mentally, emotionally and financially for the pet. Before you embark on this journey of raising your pet, it is important to evaluate your resources and make sure that you are ready.

You should also evaluate the practical side of things. It is important that you know that the cost of bringing up a ferret might be more than the cost you would have to encounter while raising a dog or a cat. It is important to have a thorough understanding about the animal. Spend some time to know everything about the ferret. This will help you know your pet better. The more you know about your pet, the better bond you will form with him. Whenever you get a pet home, you have to make sure that you are ready for the responsibilities ahead.

A pet is like a family member. This is the basic requirement to domesticate an animal. It is more than important that you take care of all the responsibilities for the animal.

If you wish to raise a ferret as a pet, there are many things that you need to keep in mind. It can get very daunting for a new owner. Because of the lack of information, you will find yourself getting confused as to what should be done and what should be avoided. You might be confused and scared. But, there is no need to feel so confused. After you learn about the ferrets, you will know how adorable they are. You should equip yourself with the right knowledge.

It is important that you understand the basic behaviour of the ferret. This will help you to understand what lies ahead of you. If you understand how a ferret should be cared for, you will make it work for yourself. You should aim at learning about the animal and then doing the right thing for him. This will help you to form a relationship with him. Once you form a relationship with the ferret, it gets better and easier for you as the owner. The pet will grow up to be friendly and adorable. He will also value the bond as much as you do. This will be good for the pet and also for you as the pet owner in the long run.

If you are in two minds whether you need a ferret or not, then this book will make it simpler for you. You should objectively look at the various advantages and disadvantages of owning a ferret. This will help you to make your decision.

If you are looking to domesticate the ferret, then you might be having many questions and doubts. You still might not be sure whether you want to buy the ferret or not. If you are still doubtful, then this book is meant to help you make a well-informed decision. This book is meant to equip you with all the knowledge that you need to have before buying the ferret and bringing it home. This book will help you understand the basic behaviour and antics of the animal. You will also learn of various tricks and tips. These tips and tricks will be a quick guide when you are looking for different ways to have fun with your pet. It is important that a prospective buyer has all the important information regarding a ferret.

You need to make sure that you are ready in terms of right preparation. This book will help you in this preparation and be a better owner for your pet.You will learn many ways to take care of your ferret. This book will try to address every question that you might have when looking at raising the ferret. You will be able to understand the pet and give it the care that it requires.

You can expect to learn the pet's basic behaviour, eating habits, shelter requirements, breeding, grooming and training techniques among other important things.

In short, the book will help you to be a better owner by learning everything about the animal. This will help you form an everlasting bond with the pet.

Chapter 1. Understanding a ferret

The word ferret has been derived from a Latin word, furittus. This word actually means little thief. This could be a reference to the ferret's behaviour of taking things and hiding them away to keep them safe and sound.

The domestic ferret belongs to the weasel family of Mustelidae. The domestic ferret is also referred to as Mustelaputoriusfuro. The weasel family has many other members along with the domestic ferret, such as otters, minks and also the black footed ferret which is an endangered species.

The ferrets are very beautiful to look at. They are small and tiny. Another interesting thing about the ferret is that they can be found in different colours, such as cinnamon, silver, white, sable and albino.

The ferrets are slowly becoming more popular as pets. They are very small in size. They weigh around 2-5 pounds. The ferret will grow up to a height of 5 inches. The length would be around 20-22 inches.

Earlier, people hesitated to domesticate these beautiful animals. The main reason for this was that not much information was available about these animals. There were too many doubts in the minds of the people, and there was no source to clear these doubts.

But now, it has become easier for people to learn more about these animals and domesticate them. The adorable ferrets are slowly becoming the most sought after pets. These animals are very cute to look at. Everybody is attracted to them.

They are very attractive and people love to keep them at home. They look very adorable and sweet. They have small bodies and are easy to handle. This is also an added advantage for them as a household pet.

If you wish to raise a ferret as a pet, there are many things that you need to keep in mind. You need to make sure that you are ready in terms of preparation to domesticate the animal.

Though many people have started domesticating these animals, you could still be confused on what needs to be done. So, it is important to be acquainted with the dos and don'ts of keeping a ferret as a pet.

It is known that ferrets are very entertaining and lovable animals. They will make a great pet. But, this does not mean that they are deprived of all that they would have found in their natural habitat.

It is very important to understand the requirements of an animal and his natural instincts. The ferrets also have certain specific requirements that need to be met. You have to understand their specific requirements before you can adopt or buy them.

1).Scientific classification

If you wish to study the classification of this animal scientifically, then the following information will help you:

Kingdom – The Animalia

Phylum – The Chordata

Class – The Mammalia

Order – The Carnivora

Family – The Mustelidae

Genus – The Mustela

Species – the M putorius

Sub species – The M p Furo

Being a member of the Mustelidae, the ferret is a close relative of the weasel. There are many studies that have proven that the basic nature of the ferret has some similarities with the wolverine, skunks and badgers.

2). Domestic ferret

The domestic ferret is not a wild animal. There is evidence that proves that these animals have been domesticated for more than a thousand years!

They are domesticated all over the world. Their characteristics help them to be great domestic pets.

Most people believe that the domestic ferret belongs to the class of rodents. But, this is not true. They are not rodents. The ferret is a descendant of European Polecats.

But, the domestic ferret will have characteristics that are unique to them and will make them different from their ancestors. For example, the ferret will enjoy social circles, which is not a characteristic that the Polecat will exhibit.

In the past, domestic ferrets were also considered as working ferrets. It is known that they are still used in the UK to guard barns. These ferrets help the owners to get rid of rodents.

Ferrets were also known to scare off animals such as rabbits from their hiding areas. This process is called ferreting. The working ferrets were also utilized for this purpose.

If you are planning to bring a ferret home, then you should know that these animals love to sleep. You might be surprised to learn that these animals can sleep for 13-15 hours a day.

The ferret feels the most energetic and active during late evening and also morning time. This is the time when the ferret might want to go out and play. You will see the ferret jumping and running because of the reserve of energy in him.

The ferret is naturally very fond of tunnelling. They will always be on the lookout to dig a tunnel. When it comes to sleeping, the animal will prefer a safe, dark and enclosed area. This gives a feeling of security to the pet.

These animals are attention seeking animals. If they are not sleeping, they will want to play with you. You should note that you can't keep them in a cage for too long.

They get depressed and stressed when they are lonely for too long. This animal will require a lot of attention. If the ferret has its way, it will only sleep and play with you rest of the time.

You will have to understand the cycle of your pet ferret so that you can take out some time from your schedule to play with them. You will have to pamper them a lot.

3). Life span

A ferret has a life span of 6-10 years. But, there have been instances where ferrets have lived longer than their average life span. The key is to provide them the right environment and also the right nutrition. This will help them to grow, stay healthy and live longer.

It is also important to note that ferrets are susceptible to various disease causing viruses and bacteria. Once a ferret gets a dangerous and life threatening disease, it can be very difficult to cure him.

The pet ferret will require you to pay a lot of attention to its health, be it vaccination or health care. The pet will definitely live longer if you make sure that you do all that is necessary for its health.

4). Ferrets in their natural habitat

In their natural habitat, the ferrets will be found together in groups. Such groups of ferrets are referred to as a business. The young ferret is often referred to as a kit. While the female ferret is often called a Jill, a male ferret is called a hob.

These ferrets are carnivorous by nature. They love their meat and dwell on a meat diet happily. It is known that previously, the ferrets would eat only meat. Their main food consisted of small animals, such as rabbits and mice.

The ferret has a very different digestive system. The digestive tract of the animal is very small. This arrangement has an effect on the food the ferret eats and the ferret's health in general.

The ferret has a high metabolism. While it can digest food faster than most animals, it can't digest many kinds of food. In its natural habitat, the animal is used to eating strictly meat.

The digestive system of the animal has accustomed to the habits of the animal for years. The digestive tract is not well equipped to digest all kinds of vegetables and plant protein.

A ferret will also have a scent gland. This will be located near the anus of the animal. The scent gland is used generally when the animal wishes to establish a territory.

They will also use it when they are scared of an impending danger. The release of the smell will be an attempt from the ferret to scare off the thing that scares it. There is a chance that your pet might not be able to use its scent gland.

Generally, the ferrets that are sold by the pet shops are neutered or spayed. Your breeder might also have neutered or spayed your ferret. You should know that if the ferret is neutered or spayed, then he would be unable to use the scent gland.

5). Teeth

The ferret is known to have sharp teeth. A person who has been bitten by a ferret will tell you this. There are basically four types of teeth in the ferrets. Different types of teeth serve different purposes for the ferret.

There are twelve teeth in the front that are called incisors. The primary purpose of these teeth is grooming. They are located between other types of teeth that are called canine and are very small in size.

The second type of teeth are called canines. These are very sharp and are basically used to tear apart the animal's prey. They have four canine teeth.

The animal uses the pre molars to chew the food that they eat. They have 12 of these teeth. These teeth can be found at the immediate back of the canines and on the side of the ferret's mouth.

The ferrets have special teeth to crush their food at the back of their mouth. These are called the molars and there are 6 in total, with four being on the bottom and two of them being on the top.

6). Neutering or spaying

When you domesticate an animal, it also concerns you to understand its breeding cycles. As an owner, you need to make a decision whether you would want your pets to have progenies or not. This should be a well thought out decision, so that you can take the required actions.

These ferrets are often neutered or spayed by the owners or the breeders. The neutered male ferret is referred to as a gib and the spayed female ferret is referred to as a sprite. Neutering or spaying is an important part of ferret domestication.

When you are sure that you don't want your female ferret to breed, it is better to spay the animal before it is too late. Similarly, neuter the male ferret if you don't want breeding. Many breeders sell ferrets after neutering or spaying the ferrets.

It should also be understood that neutering or spaying the domestic ferret will have its consequences on the ferret. It is better to understand these consequences, talk to the veterinarian about them and be prepared for them.

When a female ferret is not spayed, you will have to breed it. In case you are unable to do so, there will be health complications with the ferret. The female ferret could suffer from aplastic anaemia.

7). Breeding in ferrets

A female ferret will be sexually mature by the age of about four months. On the other hand, the male ferret will reach sexual maturity when he is six to nine months. If you wish to breed the ferrets at home, then you will have to be prepared for this task.

It will not be as simple as breeding a male animal and female animal. The ferrets can acquire diseases if the wrong parents are mated. Closely related ferrets should never be mated if you wish to keep the progeny healthy.

You will be required to get your female and male ferrets tested for their genes to establish whether they can be successfully mated or not. It is better to discuss this in detail with your vet.

If you are planning to breed your male and female ferret at home, you should be on the lookout for the signs that show that the ferrets are ready.

It is said that the first spring after a ferret's birth is the season for its mating. The days will get longer and warmer during this time. The first tell-tale sign that the ferrets are ready is that they secrete an odour and their skin become shiny because of this secretion of oil.

The female ferret (jill) should be four months of age whereas male ferrets should be at least six months of age. The female will have a discharge from her vagina and her vulva will also be swollen.

The male ferret (hob) will be ready when his testicles enlarge and are visibly dropped from his body. The hob will also use his oil and urine to mark his territory. You might even see him rolling in his own urine. Once you are sure that the jill and the hob are ready for mating, keep them in a single cage.

You should be ready to see some violent mating, which will not be a great sight. There will be a lot of dragging, wrestling and noise making. The male will bite the female over her neck so that she secretes her eggs.

The actual mating might take many hours or even many sessions. You should be prepared for this. This is normal, so there is nothing to worry or fret about.

After the process, keep them in separate cages and observe your female ferret for signs of conceiving. The first sign will be that she will gain weight and will also pull her own fur from her body and tail. She will also make noises.

But, sometimes the increase in weight can happen because of various hormones, so you should keep a check on her. She will eat more as the pregnancy approaches. An ultra sound can confirm the pregnancy. In case of a failure, you can try to mate them again.

8). Legal regulations

When you are studying the domestic ferrets closely, it is important to understand the legal regulations that govern them. This will help you to know how easy it is for you domesticate the ferret in your country.

▪ Brazil: If someone in Brazil wishes to domesticate these animals, then it is possible with a few legalities. It is imperative to get the ferrets sterilized. And, you will also be required to get the ferret a microchip identification chip.

▪ New Zealand: Since the year 2002, New Zealand has made it illegal to breed and sell these animals. If you wish to domesticate a ferret in New Zealand, you will have to go through various legalities to make this possible.

▪ Japan: Certain places in Japan make it very simple for the owners to domesticate the ferrets as there are no restrictions. But, some areas will require you to register yourself and the ferret with the local body in your area.

▪ Australia: There are some places in Australia that allow the domestication of the ferrets. The states may require you to have a legal license to domesticate them. There are some places where the keeping of these animals is illegal. For example, Queensland bans the domestication of ferrets.

▪ United States: Previously, the United States had banned the keeping and breeding of ferrets. But, as the ferrets became popular in the late eighties and early nineties, the laws were changed for many places. Many places such as

California still don't allow the domestication of ferrets. Many military bases have banned the ferrets in their areas.

You have to understand the licensing requirements in your area before you can keep a ferret. This is important so that you can avert any future issues and problems.

If you are looking to bring home a ferret, then you should contact the animal shelter in your area. It is also recommended to look for a ferret in an RSPCA rescue centre.

You should make sure that before you make the payment and buy a ferret, you understand all the legalities in your area. Your breeder will also help you to understand all the formalities.

This is one of the most important steps while looking to domesticate a ferret. You should make sure that you inform yourself well about this and speak to all the concerned and relevant people.

9). Things to know before you buy the ferret

As a prospective owner, you might be wondering about the costs that you need to prepare for. You might also be thinking as to what is so special about bringing a ferret home.

Why should you be so prepared? Why isn't it like getting any other pet? To clear the various doubts in your head, you should understand the nature of the pet and also the various costs that you will incur while raising the pet.

Your ferret will have various things that will make him unique. There will be specific requirements of the pet. For example, in regards to diet and housing, the animal has some very specific needs. You should be able to understand the needs and then also fulfil them.

As the owner and parent of the pet, you will have to make attempts to fulfil all the needs of the animal. You should also be prepared on the financial front to take care of these needs.

It is better that you plan these things well in advance. This planning will help you to avoid any kind of disappointment that you might face when there are some payments that need to be made. This will help you to be prepared and also avoid any deadlock or fix at the later stage.

Before you are all set to buy the ferret and domesticate him, it is also important that you work on all the factors that will affect the domestication of the animal. This section will help you in understanding what you can expect in terms of finances when you are planning to bring a ferret to your household.

In the very beginning, you need money to buy the ferret. Once you have spent money on buying the animal, you should be ready to spend more money. You can expect to spend money on the shelter, healthcare and food of the animal.

While there are certain purchases that are only a one off and fixed, you will also have to be prepared for unexpected purchases that you will have to face once in a while. You have to be ready to bear various other regular costs continuously over the years.

Being well prepared is the best way to go about things. There are basically two kinds of costs that you will be looking to incur, which are as follows:

***The one-time or initial costs**:* The initial costs are the ones that you will have to bear in the very beginning of the process of domestication of the animal. This will include the one-time payment that you will give to buy the animal.

There are other purchases that would come under this category. The initial costs that you will face when you have decided to domesticate a ferret are the purchasing price of the animal, the permits and the license fee, the vaccines, purchasing price of food containers and the of the enclosure.

***The regular or monthly costs**:* Even when you are done with the one-time payments, there are some other things that you won't be able to avoid. But, these finances can be planned well in advance. You can maintain a journal to keep track of them.

The monthly costs are the ones that you will have to spend each month or once in few months to raise the ferret. This category includes the costs of the food requirements and health requirements of the pet.

The various regular veterinarian visits, the sudden veterinarian visits and replacement of things come under the monthly category.

The various purchases you can expect

While you are all excited to domesticate the ferret, you should also start planning for all that you can expect in the future while raising the pet ferret. You can expect to incur the following:

Cost of buying the ferret

The initial purchasing price of the ferret could be higher when compared with the initial amount incurred in purchasing other regular animals. If spending a high amount of money is an issue with you, then you will have to think twice before purchasing this animal.

On the other hand, if spending money is not an issue then you should understand the other important factors for raising a ferret and accordingly make a decision.

If you are planning to buy a ferret from a pet shop, then you can expect to pay somewhere around $100/£77.62. In general, the ferret can be anything from $65/£50.45 to about $250/£194.05.

But this is only the purchasing price of the ferret; you will have to pay for the vaccinations of the animal also. These vaccinations could be anywhere between $100/£77.62 to $400/£310.48.

You should make sure that you get the ferret medically tested before buying it. The examination and tests will also add on to the initial price. You also have the option to adopt a ferret. This will help you to avoid the initial purchasing price, though the other costs for raising the ferret will remain essentially the same.

You should also understand that neutering and spaying the animal will additionally add on to the price. If the ferret has already been neutered and spayed, the breeder will inform you about this.

Most breeders mark the neutered or spayed ferrets by two dots near the ears. Good breeders will always make sure that the ferrets that they are selling to you and other buyers are in the prime of their health.

They will take care of their vaccines and health in general. This again means that the breeder will charge more.

If your breeder has taken care of the initial doses of vaccines, then you should be fine with paying a little extra to this breeder because he has saved you from running here and there to get these important procedures done.

Cost of shelter

When you bring a pet home, you have to make the necessary arrangements to give it a comfortable home. The shelter of the animal will be his home, so it is important that you construct the shelter according to the animal's needs.

If the pet is not indoors, he will most likely be in his cage. If the cage is not comfortable, you will see your ferret withdrawing from it. So, it is important to make this one time investment in a way that is best for the ferret.

The ferret will require a good quality and comfortable cage as its home. The ferret will spend a lot of time outdoors, but you should buy a good cage for the animal to rest and sleep.

If their shelter is not comfortable, the pet will be restless all the time. Even if you construct a very basic cage for the animal, it should have the necessary comfort.

This is a one-time cost, so you should not try to save money by putting the pet's comfort at stake. The price of shelter will depend on the type of the shelter. You can expect to spend anywhere between $50/£38.81 to $500/£388.1 for the cage of the ferret.

The cage should be accessorized well by you. The cage will require some basic stuff, such as bedding, hammocks and toys. The higher end cages will have tunnels to keep the pet occupied and happy.

These are the extra things that you will have to incur in addition to the basic price of the cage. This should be around $70/£54.33. Though they might not be necessary, these accessories will make the cage fun for the ferret.

Cost of food

A domesticated ferret will mostly be fed meat and cat and kitten food. Ferrets are carnivorous animals, so it is important that they are served more meats in addition to the basic kitten food. You might also have to include various supplements to give your pet overall nourishment.

This is a basic requirement of the pet that you can't evade. It is important that you understand the food requirements of your ferret in the beginning, so that you can be prepared on the monetary front.

This is important because if the animal does not get all the appropriate nutrients in the right amount, his health will suffer, which again will be an extra cost for you. So, make sure that you provide all the necessary nutrients to your pet animal.

You should be prepared to spend about $20/£15.52 to $40/£31.05 on the diet of your pet every month. This will vary depending on various factors, such as the brand of products that you choose and also your exact location.

The kinds of food that you feed your pet will also affect the exact amount food that you encounter per month. You should remember that the more lax you are regarding the money that goes into food, the higher the vet costs will be!

If your pet is well fed, it will not fall sick that often. This will automatically reduce the amount of money that you would have to spend on the veterinarian and medication.

Cost of health care

It is important to invest in the health of a pet animal. This is necessary because an unhealthy animal is the breeding ground of many other diseases in the home. Your pet might pass on the diseases to other pets if not treated on time. This means danger for the pets and also the members of the family.

You will have to take the ferret to the veterinarian for regular visits on his health. He will be able to guide you regarding any medications and vaccines that the pet may need.

It is advised that for the very first year of domestication, you are extra careful regarding the health of the animal.

You should also be prepared for unexpected costs, such as sudden illness or an accident in the ferret. Health care is provided at different prices in different areas. So, the veterinarian in your area could be costlier than the veterinarian in the nearby town.

You should work out all these things right in the beginning, so that you don't suffer any problems later. Realizing at a later stage that you can't keep the animal and giving it up is never a good idea.

Ferrets can get sick very easily. You will have to invest in their health care. They are susceptible to many diseases such as respiratory infections and Adrenal diseases.

You should understand that taking care of these animals will require special skills. You should make sure that the veterinarian that you consult for your pet ferret is experienced in handling such animals.

You should also be prepared to spend more money on their health than what you would have spent on other pet animals. It is believed that you should have an extra $1000/£776.2 saved for your ferret's emergencies. He might require an operation or surgery because of a disease.

You will have to spend money on getting the vaccines for the ferret. These vaccines are critical to save the pet from diseases and deficiencies at a later stage, so make sure that you don't miss them.

The ferret will require vaccines against rabies and some other diseases. There are many breeders that take care of the early vaccines of the pet before giving it to the new owners.

You should talk to your breeder regarding this. The breeder might include the money spent on vaccines in the final price that he might charge for the animal. You should also keep a track of the vaccines, so that you don't miss any.

Cost of hygiene

A pet needs to be clean and hygienic. If you fail at maintaining hygiene levels for your pet, it will only lead to other complications. The entire household will be affected if the pet is not clean.

Germs travel very fast, and before you realize there will be may hygiene related issues in your entire house.

There are many owners that insist on litter training. If you too wish to litter train your pet ferret, you will have to buy the required products for the same. You should invest in a high quality cat shampoo. This will be enough for the baths of the ferrets.

You will also be required to invest in a good detergent, which could be bleach. This will be needed to clean the cage and all the areas where the ferret might defecate.

Apart from the detergent, you should invest in some good cleaning products and sanitizers. This will be necessary because the smell of the faeces will be difficult to get rid of. You will need these cleaning products to eliminate the smell of faeces from the floor or from the cage area.

On average, you can look at spending $15/£11.64 to $30/£23.29 a month on the hygiene requirements of your pet. This is essential and important so that the surroundings of the ferret can be kept clean.

Other costs

Although the main costs that you will encounter while raising your pet have already been discussed, there will be some extra things that you will have to take care of. Most of these are one-time costs only.

You will have to spend money to buy stuff such as ferret bedding, accessories, food and water bowls and toys for the pet.

You can expect to spend some $200/£155.24 on these things. The exact amount will depend on the wear and tear and the quality of the products. In order to keep a track of things, you should regularly check the various items in the cage of the pet.

If you think that something needs to be repaired or replaced, you should go ahead and do it.

10). Ferret insurance

You can get insurance for your pet ferret. In fact, you should plan this well in advance. This insurance will help you to take care of the vet bills and surgery and injury costs. A healthy pet is like an asset, but when the same pet falls ill, it can be a nightmare.

The ferrets have a tendency of getting sick very easily. It is known that by the time the disease is confirmed, the ferret needs to be operated on. They also get injured. These procedures can cost you thousands of pounds and dollars.

If you buy an insurance to cover these conditions, you will save yourself from a lot of trouble. Depending on the insurance you buy, you can also cover the cost to neuter or spay and also the regular clinic visits. There are some companies that will give you discounts on clinic visits.

There are many kinds of wellness packages for the ferrets. For example, you can buy a package to cover all the vaccines throughout the life of the ferret.

Insurance for dogs and cats is very common, but insurance covers for unusual pets are not that common. But, there are some companies that can help you with ferret insurance, such as Exotic direct in the UK and Pet assure in the USA.

These companies have different kinds of insurance. You can choose according to your requirement. You can also get a package deal if you are looking to insure more than one ferret.

It is very important to understand the policy that you are buying. You should know which diseases are covered in the policy. There are some policies that provide security against only a few ferret diseases.

Make sure that you discuss the policy with the vet. The company that you choose to buy the insurance with will also take into account the health of your ferret. Each insurance policy will have a set of conditions that will have to be met.

When you buy insurance, you have to pay a deductible amount and regular premiums.

You can expect to pay a deductible amount of about $100/£77.62. The first $100/£77.62 spent on the health of the ferret will have to be paid by you. After that amount has been spent by you, the insurance cover will take care of the rest.

You will also be required to pay premiums that need to be paid regularly to keep the insurance policy active. The premium that you will pay will depend on the kind of insurance policy you opt for.

11). Maintaining hygiene

You should also note that it is not enough to buy a pet and provide him with a cage or food. You also have to understand the hygiene requirements of the pet ferret.

Hygiene is always an important factor when you are keeping a pet. The pet's surroundings have to be as clean as possible. This is important so that you can keep your pet away from various diseases.

You will have to make sure that the cage and the surroundings of the pet are clean. You have to make sure that you have the time or manpower to get the cleaning done.

Your pet might defecate on the carpet or floor. While you can litter train the pet, you also have to be prepared for such things. An attempt has been made to cover all the necessary issues that you will encounter when planning to take care of the ferret's hygiene.

When you are keeping a ferret, hygiene is all the more important. You should make sure that you don't fail to meet the standards that are required to keep the pet healthy.

The habitat of the pet should be as clean as possible because a dirty environment will only lead to germs and diseases. Keep the cage clean at all times, without exception.

You have to make sure that your hands are clean before you can feed the ferret. When the ferret is sick, you might have to hand feed him. This point becomes all the more important at that time.

Like you would wash your hands before and after eating food, you should maintain a routine of washing your hands nicely with soap before and also after feeding the pet.

When you are looking to maintain hygiene for your pet, you should understand that there are certain tasks that you will have to do once a week or once in fifteen days.

There will also be certain cleanliness related tasks that you will have to do every day, such as keeping the litter boxes clean. You can't postpone these tasks to the next day.

You should make sure that these tasks are done on time and that you dedicate yourself to get these tasks done each time. If you fail to do so, the ferret will have to suffer.

12). Bringing home a healthy ferret

A major concern that many prospective owners and buyers of the ferret have is how to make sure that the animal that they are getting home is healthy.

It is extremely critical that you get a healthy ferret to your home because once you get an unhealthy kit you will only make things worse for yourself and the pet.

In the excitement of getting a new pet, you shouldn't forget the basic checks that you need to do before bringing the ferret home. The last thing that anyone would expect after finding a breeder and getting an animal is that it is not in good health.

You will not know how to care for the sick pet. The pet's health will deteriorate. You will be spending thousands of dollars just on the health of the animal.

The following pointers will help you to make sure that your future pet is in the prime of its health:

- It is very important to bring a healthy pet to your home. You should definitely avoid bringing an injured animal home.

- Make sure that you learn as much as possible about the ferret before you decide to buy him and bring him to your home.

- Even if the animal has had health issues in the past, it can be a matter of concern for you.

- A younger ferret will have issues that could be different from an older ferret. You need to make sure that you understand all these issues in detail.

- If you are buying an older ferret, you need to be all the more vigilant because they could carry some infections. First and foremost, you should check the health care card of the animal that you wish to buy.

- All good breeders will maintain a health card, which will have all the details of past diseases and infections. This health card will also help you to understand the vaccine cycle of the animal.

- You will be able to understand which vaccines have been completed and which ones are due.

- It is always better to buy a ferret whose vaccines have not been cancelled or missed.

- It is important that you closely examine your prospective pet. You should look for any abrasions on his skin.

- His skin should not be torn or bruised.

- You should make it a point to check the body temperature of the ferret. The body temperature should be normal.

- If the ferret is too cold or too hot to touch, then there is some problem with its health.

- You should closely look for any kind of injuries. If you find anything that does not seem normal, then you need to discuss it with the breeder.

- The ferret should not have any broken limbs. You should be able to check this manually.

- You should look for any hanging limbs. A hanging head or limb could mean that the pet is severely injured.

- Also, carefully inspect the tail and stomach area. There should be no abrasions.

- It is important that the ferret is devoid of any infections or diseases when you bring him home.

- It is advisable to take the help of a qualified veterinarian to be sure of the kit's health conditions.

- He will be the best judge of his condition. A good vet will always guide you in the right direction for the ferret.
- You should discuss at length about the concerns that you have regarding the ferret.
- You should follow all the instructions that the vet gives you because they will be for the benefit of the animal.
- You should only keep the ferret if you are convinced that you will be able to care for the little animal.
- After you have brought the ferret home, you should keep him isolated to keep an eye on him.
- You should allow him inside the house only after a few days of checking if everything is normal with the ferret.
- In case of any issues, you should consult the vet and the breeder.

Chapter 2. Owning a ferret

If you wish to own a ferret or even if you already own one, it is important to understand the basic characteristics of the animal. You should know what you can expect from the animal and what you can't.

This will help you to tweak the way you behave with the ferret in the household, which in turn will help to build a strong bond between the ferret and you.

A pet is like a family member. You will be more like a parent than like a master to the pet. You will be amazed to see how much love and affection your ferret will give through his ways and actions.

You have to make sure that the animal is taken care of. The animal should be loved in your household. If your family is not welcoming enough for the pet, the animal will lose its sense of being very quickly.

If the ferret does not feel wanted and loved in your home, you will see a decline not just in its behaviour but also its health. This is the last thing that you should do to an animal. An animal deserves as much love and protection as a human being.

You should be able to provide the pet a safe and sound home. Your family should be caring towards the pet. You have to be like a parent to the ferret. This is the basic requirement when planning to bring an animal home.

All these points are not being discussed to frighten or scare you. In fact, these points are being discussed to make you understand that you have to know the right ways to domesticate a ferret.

Ferrets are known to be very loyal animals. If they establish a trust factor with you, they will always remain loyal to you. This is a great quality to have in a domesticated animal.

Along with being loyal, they are also known to possess great intelligence. They will actually surprise you with their intelligence. This makes the pet all the more endearing.

When the ferret is in a happy mood, he will jump around the entire space. His unique ways and antics will leave you and the entire family in splits. If you have had a bad day, your pet will surely help you to release all the tension and enjoy life.

They are also very entertaining and playful. You can expect the entire household to be entertained by the unique gimmicks and pranks of the ferret. If you are looking for a pet that is affectionate, lovable and fun, then the ferret is the ideal choice for you as they won't disappoint you.

In spite of all the qualities of the ferret, it is often termed as a high maintenance pet. If you are still contemplating whether you wish to buy a ferret or not, then it is important that you understand all about the maintenance of the pet, so that you can make the right choice for yourself.

1). Advantages and disadvantages of domesticating Ferrets

If you are in two minds whether you need a ferret or not, then this section will make it simpler for you. You should objectively look at the various advantages and disadvantages of owning a ferret. This will help you to make your decision.

It is important that you understand that owning any pet will have its advantages and disadvantages. You should see whether with all its pros and cons, the animal fits well into your household.

A few advantages and disadvantages of domestication of ferrets have been discussed in this section. If you are a prospective buyer, then this section will help you to make a wise decision.

There are people who are impressed by the adorable looks of the ferret. They think that this reason is enough to domesticate the animal. They believe that just because the pet is tiny, it wouldn't require any maintenance. But, this is not true.

Domestication of a ferret has its unique challenges and issues. If you are not ready for these challenges, then you are not ready to domesticate the animal. Once you understand the areas that would require extra work from your side, you will automatically give your very best in those areas.

If you have already bought or adopted a ferret, then this section will still help you. The list of pros and cons of ferrets will help you to prepare yourself for the challenges that lie ahead of you. This list will help you to be a better parent to the pet and to form an ever-lasting bond with your beloved pet.

Advantages of domesticating a ferret:

If you are still not sure about adopting or buying a ferret, then you should know that there are many pros of domesticating a ferret. They are loved by their owners and their families because of some amazing qualities that they possess.

This animal can definitely prove to be a great pet for your household and your family.

The various advantages of domesticating a ferret are as follows:

- The size of the ferret makes it an ideal choice as a pet. They weigh between 7 to 10 pounds, which makes them very light in comparison to many other commonly domesticated animals.

- Their looks make them adorable and cute to look at. They are loved by one and all. Who wouldn't want to have a pet that is beautiful to look at?

- People who love pets that can be lifted and cuddled will love the ferret. A ferret will allow you to lift it and play with it.

- This pet will be the centre of affection for all the family members and also for each and every visitor of the house.

- Ferrets are known to be very loyal animals. They will love your presence around them and will show you that they love you by their own unique ways.

- If they establish a trust factor with you, they will always remain loyal to you. Loyalty is a very good trait in an animal. This is a great quality to have in a domesticated animal.

- Ferrets are also known to possess great intelligence. You should be prepared to witness their intelligent antics and gimmicks.

- A ferret is a very sharp animal. It is always good to have a pet that is intelligent and sharp.

- These animals don't require a regular walk like many other pet animals.

- If you care well for the pet, he will also respond in a very positive way. When the ferret is in a happy mood, he will jump around the entire space. His unique ways will keep you happy and entertained.

- If there are kids in your home, then they will fall in love with this pet. But, you should monitor the interaction of the kids with the pet. This is important to keep everyone safe and sound.

- The ferret is a very active and energetic animal when it is awake. It will keep itself happy and entertained. You will not have to worry much about the pet.

- These pets sleep a lot. They will sleep most of the time, which gives you a lot of free time to do whatever you want to do.

- One of your main concerns could be the diet of the pet. Even if you love your pet dearly, you would want to avoid any hassles while feeding the pet. You might not have the time to prepare special food all the time. In case you domesticate a ferret, then you will not have to worry too much about the diet. The ferret can be served meat with some store bought ferret food or kitten food. There are easily available food items to ensure that the right nutrition is given to your pet.

- The ferrets don't overeat, so you don't have to worry in this aspect. You can leave food in the container and the ferret will eat as much as is required. They are used to eating several small meals.

- The ferrets can be trained against nipping. You can also litter train them. You can teach them some easy tricks to have more fun with them.

- A very important point to note here is that their demeanour will depend a lot on how they are raised. The preparation has to begin right from the start. You can't expect them to suddenly become friendly after years of wrong treatment. If they are raised to be social, they will be very social.

- The animal will secrete a gland that will make him smell when he is frightened and alarmed. But, if the pet is neutered or spayed, there will be no stinking and dirty smells. This can be a great relief for many people because who wants a pet who smells all the time?

- Ferrets live in groups in their natural environment. This makes them tolerant towards other ferrets. The ferrets will wrestle and play with each other. There is a very slight possibility that they will not get along, but if you want more than one pet, the ferret is for you!

- The ferrets have a very good sense of smell. This quality helps them in evading danger and also helps them to get trained.

- Ferrets have a fairly long life, if they are taken care of. You can make a strong emotional bond with your pet and can enjoy the fruits of the bond for years to come.

Disadvantages of domesticating a ferret:

While you have studied the advantages of domesticating a ferret, it is also important to learn about the various disadvantages that come along with them. Everything that has merits will also have some demerits, and you should be prepared for this.

The adorable and friendly animal has his own set of challenges when it comes to domesticating them. It is important to understand these disadvantages so that you can be better prepared for them. Following are the disadvantages of raising a ferret:

- The ferrets are very energetic by nature. This behaviour could be difficult for a first time owner.

- These animals have a very unique temperament, and it would require patience from your side to understand this kind of temperament.

- These animals sleep a lot. This can be an advantage for you, but if you want a pet that will play with you all day, then you are in for a loss. The ferret will play with you when it wants to.

- The ferrets can't see very well. The ferret can get hurt if it is too dark because of their weak eye sight.

- These animals are known to eat smaller but very frequent meals. So, you have to make sure that the ferret has something to eat every 2-3 hours. Such maintenance could be difficult for some people.

- They get sick very easily. A lot of care has to be taken to ensure that they maintain good health.

- They catch disease causing bacteria and viruses very easily. Once infected, it is difficult to treat them.

- The ferret has a habit of nipping. This is a habit that helps them to interact with other ferrets, but they can nip you also. Nipping can hurt you, but you can train the ferret against such behaviour.

- The food that you serve them might be easily available, but the right brands might not be too cheap.

- They are definitely not suitable for someone who is looking for a quiet and calm pet. They are energetic, will run around and will also make noises.

- The ferret can run from one place to another in a matter of seconds. A new owner might find it very difficult to keep track of this pet due to his over enthusiastic and energetic nature.

- Because of his energy levels, the ferret can run into things and can get hurt very easily.

- The ferrets are known to have very sharp teeth that can hurt small children. So, you need to make sure that the children are never left alone with the pet.

- They are so small that you can lose them if you don't keep track of what they are doing. This is very important.

- If the ferret is lost, it is almost impossible to find it. He will not be able to find his way back to the house. And, the ferret is so small that anything can happen to him when he is lost.

- The ferret has a tendency to run into danger every now and then. You have to be very serious about ferret proofing the house, else you will lose him.

- The ferret loves to dig in the ground. In fact, the ferret will try to dig every where. If the ferret is left on its own, he might try to dig in your sofas, beds, etc.

- The pet also loves to chew. They can chew away all your furniture and bedding.

- The animal seeks a lot of attention. The ferret is a kind of pet that will require you to pamper him a lot.

- The pet can get stressed and depressed if he is left alone for long durations. You can't leave him in the cage for too long.

- The cost that you will incur while buying and raising is higher when compared to other pets, such as the dog and the cat.

- You will have to spend a lot of money on the vaccination and healthcare of the ferret.

- If spending too much money is an issue with you, then you will have to think twice before purchasing the animal.

- These animals love playing and running around. These pets are fond of exploring things. They can create a mess if not monitored.

2). Introducing the ferret to a new ferret

There are many people who want to keep more than one ferret at their home, but are scared of the consequences. What if the ferrets don't get along? What if the older ferret gets insecure? If you too are sailing in the same boat and are having the same set of questions, then this particular section will help to clear most of your doubts.

To begin with, you should know that keeping more than one ferret in a household is not a bad idea. In most cases, the ferrets keep each other company. A ferret can get severely stressed of he is lonely. If there are two or more ferrets in the home, the ferrets are likely to be very happy.

Ferrets are used to being with each other in their natural surroundings. They get along well with one another. They'll fight and play and will keep each other entertained.

So, if you are having doubts about bringing home another ferret, then you need not be scared. But, you have to be careful while introducing the ferrets to each other.

If you have made a decision to buy a new ferret at a later stage, you will have to take certain precautions. It is always better if you get young kits of about the same age if you are sure about keeping more than one ferret. These kits will grow up together and it will be easier for them to get along.

If you have a ferret that is very old and sick, it can be very difficult for him to adjust with a new ferret. His health condition and age will make it very difficult for him to adjust to the new life situations.

On the other hand, if your ferret is healthy, you can introduce him to a new ferret by taking certain precautions.

If you give some time to the ferrets, they will get along with the passage of time. And, this will also give you some time to understand how the ferrets are adjusting to each other. There are a basic set of steps to make it easier for the ferrets.

If you think that you can bring a new ferret and just leave him with the older ferret, then it will get very difficult for you. The older pet could have territorial issues that you will have to address. Also, the new ferret could carry some disease causing bacteria or virus. You will have to make sure that the new ferret does not pass these on to the older one.

The first thing that you need to do when you are planning to bring a new ferret home is to get another cage for the ferret. You can get a simple cage for him because this will only be his temporary home. You should keep the new ferret in this cage for a few weeks.

Keeping the new ferret in an isolated cage is important for both the old and the new ferret. The new ferret will take some time to get adjusted to the new environment around him. If he is a kit, he will need some space and time to understand his surroundings.

The old ferret, on the other hand, will be protected from any disease that the new ferret might be carrying.

If the two ferrets are allowed to interact in the very beginning, this will only mean that the older bacteria and virus get transferred and the older ferret could get sick.

It is important that you keep the cage of the new ferret away from the old ferret. They should ideally be in separate rooms. It is critical that you observe a quarantine period of at least two weeks. These two weeks will help you to establish whether your new pet ferret is healthy or not.

It is important that the new ferret has had all his vaccines on time. Before you bring him home, you should make sure that you check the health card of the ferret. This will help you to understand the health condition of the new pet that you want to bring home.

It is known that the ferrets can transmit ECE to each other. There is nothing much you can do in this case, but you can definitely prevent your ferret from acquiring a disease from the new ferret. With the passage of time, if you witness your new pet to be unwell, you should take him to the vet.

After the first few weeks, you can be sure that the new ferret is not carrying any disease. Now, it is time to slowly introduce the pets to each other. But, this has to be done in stages.

If you think that the two of them can be put together and they'll suddenly become the best of buddies, then this is not going to happen.

You should understand that the older ferret is already used to a certain life style. He is used to a way of living and also to the people and pets around him. If suddenly, a new pet walks in, it will get difficult for him at various levels. He will try to fight it out with the new ferret to establish his supremacy.

This fight for survival and supremacy will not only strain the relationship of the pets, but will also lead to a lot of stress in their individual lives. Ferrets can get stressed by such situations.

As the owner, it is your responsibility to avoid any such situation where your pets have to go through unnecessary pressure and stress.

To make the transition easier for both the ferrets, you should begin by introducing them to each other's smell. After the first few weeks of isolating them, bring their cages closer. You can place the cage of the new ferret next to the cage of the older one in way that they can see each other and smell each other.

It is important to keep the ferrets in this arrangement for a few days so that they get used to each other's smell. Another trick that you apply is that you can exchange the bedding of the two ferrets after a few days. This is a technique that will help them to get identify each other's smell.

Once the ferrets are okay with each other's smell, you can be sure that it'll get better from there. Give them a few days, and during this time don't bring them together physically in one area. Just keep them around in separate cages.

Once you see them acknowledging each other, you know it is time to introduce them in a more intimate way. You can now bring them together in an open space. By now, they anyways know of each other's presence in the house.

You will find them smelling and sniffing one another. This is the ferrets' way of introducing themselves. You need to supervise such meetings. They will not harm each other, but you shouldn't leave them alone. Do this for a few days.

The introduction period could extend up to many days and weeks. You might also find your ferrets wrestling with each other. This, again, is a very normal behaviour in ferrets, so you shouldn't be surprised. Wrestling is a way to establish who the dominant one in the pack is.

They will nip each other, drag each other and flip each other. This would be accompanied by hissing sounds. You should not interfere in the wrestling bout, unless things get very serious. If you see a ferret bleeding or being too stressed and scared, then you know that it is time to intervene.

You should also know that the age of the pets will also determine as to how well they get along. It is always easier for younger ferrets to get along with each other. If the older ferret at your home has been there for many years, he will take a lot of time to accept the new ferret.

If the new ferret is very young, you can wait for a few months before you introduce him to the older ferret. This is the best for both of them. You can expect your pets to get jealous when they see you spending more time with the other one.

To balance things out, you should try to spend time with both and should also be very loving and encouraging in your words. They will sense your tone and love. You have to do this so that the ferrets don't feel lonely and left out.

A simple trick that you can make use of to help your older ferret to like the new ferret is to put 2-4 drops of Ferretone on the new ferret. You can rub this over the ferret's head and toes. The older one will lick the new one because of the smell of Ferretone.

Now, reverse the process and rub some Ferretone on the older one and the new ferret will lick. Do this simple trick for a few days. This will help the ferrets to get used to each other. You can also come up with different games that encourage you and the ferrets to spend time with each other.

If the ferrets seem to enjoy playing together, they will slowly start getting along. A strong indication that the pets are comfortable with each other is when they are curling up together. You might find them all curled up and sleeping next to each other. This is when you can put the ferrets together in a cage.

But, even this task should be done in smaller steps. Initially, keep them together only for some time. Then, slowly increase the time. Add new toys to the cage and observe the ferrets' reactions. Remove the toys that seem to create tension between the two ferrets.

In most cases, the ferrets will eventually get along. Ferrets will slowly get to like each other and enjoy each other's company. It is a matter of time before this happens.

But, if your ferrets are brutal towards each other and refuse to get along even after multiple trials, then you know that they need to be away from each other. There is a slight chance for this, but still you need to be prepared for this.

If you notice blood shed each time they are together, then this is not a good sign. While wrestling, screaming and pulling each other is fine, blood shed means that things are very serious.

You will have to step in in such extreme cases. In such a case where you feel that the ferrets are not getting along, you are advised to have two separate cages for the ferrets.

3). Ferret vaccinations

Vaccinating a pet against various diseases is an important part of pet care. You should understand the various diseases that can affect the ferret, and which ones have vaccinations.

It is important that you understand everything about ferret vaccination. This will help you keep all the vaccination doses on time. Make sure that you never miss a dose.

The breeder should also give you the health update of the ferret when you purchase him. You should know every little detail about your pet. Never buy a ferret that you are unsure about. The breeder should inform you everything regarding his vaccines.

The ferrets can get sick very easily, this makes it all the more important to do everything that would help them to be healthy. Your pet is at a high risk of canine distemper. It is extremely important to vaccinate the ferret to eliminate this danger. This should be done when the ferret is at the age of about eight weeks.

After this dose, it is important to give the ferret a dose at eleven months and also fourteen weeks. Once this is done, you will have to make sure that the pet gets a dose every year.

This is very important to keep the pet ferret safe. It is very important that you discuss the vaccination type with the veterinarian. You should ask which vaccines should be given to the ferret.

The ferret should only be given a pure canine distemper vaccine. A pure vaccine will contain the vaccine for only the specified disease. There will be no other vaccines for other conditions. It is important that the ferret gets a pure vaccine for this condition. FerVac-D is known to be a pure vaccine, and this also has been certified and approved by USDA.

Just like a vaccine against canine distemper is needed by the ferret, a vaccine against rabies is also essential. The first dose for the vaccine for the same should be given to the ferret at the age of about sixteen weeks. After this initial dose, you should make sure that a dose for the vaccine is given once every year.

Make sure that you don't miss any doses for any vaccine that the ferret requires. If your pet does not get the vaccines on time, there is a chance that he might acquire the disease. If your pet acquires rabies and the vaccines are not given to the pet, the state might decide to have the pet ferret killed.

After your pet has been given a vaccine dose, he might feel a little lethargic and would want to rest and sleep. This is absolutely normal behaviour, so you should allow your ferret to rest. If you see any abnormal behaviour or the ferret getting sick, then you should take the ferret to the veterinarian.

It is also important to keep some kind of gap between the two kinds of vaccines. You should keep a gap because this will prevent any negative reaction of the doses, and it will also be easier for the pet ferret.

You should discuss this point with the veterinarian and should plan the dates of various vaccines accordingly. This is important so that the ferret can remain in good health.

Your ferret can also be at the risk of heartworm that is caused by mosquitoes. Ferrets spend very little time in open spaces and mostly stay indoors. But, there is no way to ensure that your home has no mosquitos.

If a mosquito bites the pet, and if the mosquito is the carrier of the infection, then your ferret is at the risk of heartworm. This can be very deadly and you might not be able to cure the ferret.

The best way to prevent this disease is to get the ferret vaccinated against the infection. If you live in an area that has a lot of mosquitoes, then it becomes all the more important to get this vaccination.

You should discuss your concerns with the veterinarian and ask him for the vaccination doses and details. Depending on the area and the prevalence of mosquitoes in the area, the doses will be decided.

You might even have to get a prevention dose every month for your pet. But, this is better than getting the infection because once the ferret gets the infection it will be difficult to treat him.

4). Other important points

If you are worried whether the ferret will be able to form some bonds in your home or not, then you shouldn't worry too much about it. Focus on the right things and the rest will only follow.

Like with other pets, once you form a personal bond with the ferret, it gets better and easier for you as the owner. If you get a young ferret home and spend your time and energy to raise him, you will notice as the animal grows older, it gets very fond of you.

You will be surprised to see how loyal and friendly your pet will be towards you. The pet will grow up to be friendly and affectionate. You can also form a good bond with an older ferret.

The ferret is an animal that is known to have loads of energy. This will make him an active pet. You should lay emphasis on the quality of time you give your ferret.

As the animal grows, he will value the bond that will be formed between the two of you. It is important that you make the right efforts for this bond to be formed between the ferret and you.

If you have other pets, such as cats and dogs in the house, then this could be a very important concern for you. If you are looking to get a new ferret, then also the compatibility issues between the pet and the ferret would be on the top of your mind.

It is important that the pets in the house get along with each other. This will make the atmosphere of the house very lively. You should look for the reactions that the pets have when they are with each other.

If the pets don't gel with each other, then it is not good for all the animals. As the owner, things will get difficult for you also because the primary concern of the owner is to give protection and safety to the pets.

You might have to keep them separate to make sure that everything in the house is fine. This is fine as long as they are not disturbing each other.

It is important that you understand the various criteria that will affect the compatibility of various pets with each other. These criteria will be of the temperament of the animals and also their age.

Sometimes, the animals don't intend to harm each other. But, they don't know how to behave with each other. In this confusion, they can end up harming each other.

So, as the owner you have to be on guard to make sure that all the animals are safe and none are harmed.

Chapter 3. Decoding the ferret's behaviour

A ferret is a small, naughty animal that will keep you busy and entertained by all its unique antics and mischiefs. It is said that each animal is different from the other. Each one will have some traits that are unique to him.

While you will learn about all the unique traits that your particular ferret has by experiencing them and spending time with him, there are some traits that almost all ferrets will exhibit.

It is beneficial to know of these traits so that you are not taken off guard. You will be able to understand what is normal for this animal and what is not. This will help you to be more prepared and not be confused every time something happens.

It is important to understand the behaviour and temperament of the specific animal that you wish to domesticate. This will help you to be a better master. Your ferret might still have some surprises for you, but it is better to know of the general behaviour of the animal.

Understanding the behaviour will also help you to understand the ferret's behaviour with other animals. You will be able to understand whether your ferret will be friendly with other pets in your home.

It is important that you understand the various criteria that will affect the compatibility of various pets with each other. The given points will help you to understand factors on which the compatibility of various pets depends. These points will help you to plan how you can keep your different kinds of pets together.

The following personality traits will help you to be better prepared for your pet ferret:

1). Wrestling

If you are planning to domesticate more than one ferret, then this is one antic that you will notice a lot in your ferrets. Ferrets love to wrestle with each other. Actually, ferrets are very joyful and playful. But, they can be very rough with each other while playing. It might appear more like they are wrestling than playing.

The dominant ferret will try his antics on the poor submissive one. You will find them making sounds of excitement. Don't worry because this is a very normal behaviour in ferrets. But, do keep an eye on the ferrets so that things don't go out of hand.

2). Alligator roll of the ferret

This is a special kind of roll in which a ferret holds another ferret by his loose skin and rolls him over. This is a very common trait in ferrets. If you are planning to have more than one ferret at your home, you can expect to witness a lot of alligator rolls.

This technique is basically a way of establishing supremacy for a ferret. A strong and dominant ferret will get hold of the loose skin at the back of the submissive ferret and roll him. There is nothing to worry about because such flips are generally not harmful and the ferrets are just being playful.

If a ferret is alone, he might just flip and roll on the floor to show that he is happy and playful. You should worry if the ferret tries to try this on you. He might try to get hold of loose skin on you and might try to nip you and bite you. This can be painful for you. But, you should know that this is a very normal thing for a ferret and that you can nip train him.

3). Spotting a frightened and scared ferret

It is easy to spot a scared ferret. If you see your ferret being pushed up in a corner, you should know that something is not right. If he is making noises similar to a hiss then you should know that he is definitely scared.

It is important to know when your pet is scared so that you can comfort him and make him feel better. There are different ways to comfort different kinds of animals. For your ferret, you just need to let him be.

Leave the pet alone, but be in the vicinity so that you can keep a check on him. The ferret is known to recover on its own. He will know when the danger that frightened him is no longer there, and this will help him to get back to normal. If you really wish to help then speak a few kind words to your pet.

Don't make the mistake of taking him in your arms. This is not required right now. Just speak a few kind words to reinforce the fact that everything is fine and then just leave the ferret for some time. This works for a scared ferret.

It should also be noted here that there is another reason for a ferret to curl into a corner. This could also mean that the ferret wants to defecate. If you see your ferret tucked up into a corner, look out for the possibility of whether he is scared or not. If you don't hear him making noises or acting all scared, then this means that he wants to defecate.

As soon as you have established that this is the reason for your pet to be in the corner, you should take him to a litter box. It should be noted that if you don't act immediately, your pet will defecate on your floor or carpet. He will not wait for you to act, so you need to be quick.

4). Puffy tail

As explained earlier, a puffy tail could mean that the ferret is scared. A ferret that has taken up a corner, has a puffy tail and is making noises is definitely scared. But, there could also be other reasons for a puffy tail. A puffy tail would signify that the ferret is too excited.

If you see your ferret jumping around having a puffy tail, you should immediately know that the pet is very excited. This could happen if the pet is some area for the very first time. This could also happen if the pet is involved in a very high intensity play. Some ferrets get the puff tail when they are let out during cool morning hours.

The puffy tail is also called the 'bottle brush tail'. The tail will basically look all puffed up. You can expect your ferret to be all charged up like a rocket at this time. You will notice immense amounts of energy in him. He will dart from one end to another in seconds and you should just forget about catching up with him.

5). Dance of joy

Dance of joy is a very popular and endearing trait of the ferret. When a ferret is all excited, he will jump and flip. He will do everything that sounds crazy. He will dart from one side to another. You might see him jumping from the top of furniture. This is the dance of joy of the ferret.

The ferret might also emit certain kinds of sounds during this dance of joy. This is a simple indication for you that the pet is very excited and happy and wants to play and have some fun with you.

A new owner might not take well to this unique way of displaying excitement. There are many owners and their family members who get scared after seeing the ferret like this. But, you should know that this behaviour is completely normal and the ferret will not harm you. He is just having fun and wants to include you in his fun.

6). War dance

The war dance is basically a set of movements that the ferret displays when he is upset, frightened or angry. The movements could seem quite similar to the dance of joy to many people, but if you care to look closely, you will understand that the two are very different.

The key is to observe the body language of the pet ferret. If the ferret is sticking to a corner, is making noises, has a bent or arched back and has his fur standing, then the ferret is definitely upset, angry or scared.

If the pet is not neutered, he will also emit a very filthy smell. This should be a clear indication that something is not right with the pet. The ferret will pose like he is very strong and will be sticky to ward off the danger.

Leave the pet alone, but be in the vicinity so that you can keep a check on him. The ferret is known to recover on its own. He will know when the danger is no longer there, and this will help him to get back to normal. You have to closely observe your pet to understand his moods and the accompanying sounds.

When you see a sad or angry ferret, your first instinct as the parent would be to pick the pet up and calm him down. If you get over protective, he might just get irritated and might just bite you in frustration. You should say some kind and soothing words to the pet. When a pet ferret is angry, he needs some time to calm down.

7). Game of chase

Game of chase is one of the most favourite plays of the ferrets. If you have more than one ferret, you will often see them chasing each other. The pet will also try to get the attention of the owner either to chase him or to get chased by him.

Many new owners get scared when they see their pet all excited running around them. There is nothing to worry about because the pet is only trying to start a game of chase with you. You can take a long cloth such as a blanket or towel and run with it. Your ferret will happily run after you.

If you are the one who is chasing the ferret, then make sure that you maintain a good distance between you and the animal because these animals have another unique trait. The ferret might just suddenly stop while running. If you are not careful, you might just trample your pet ferret, which is the last thing that you would want to do.

8). Digging of food

There are certain traits that are innate in an animal. No matter how unwanted and troublesome that might seem to you, you have to get used to it. One such trait in the ferret is digging. Ferrets are very fond of digging.

While it is okay if they try to dig a blanket or the ground in the open yard, it can be troublesome if they try to dig their food. When you serve food to them in a food bowl, they will get all excited and will try to dig the food. This will not result in anything but a lot of mess.

As an owner, you can just ignore this habit of the pet ferret because it comes naturally to him. And, the ferret will eat the food that has fallen down from the food bowl. Many people might get disturbed by this habit of the pet. But, it is better to just get used to it as soon as possible.

9). Tipping of bowls

Another habit of the pet ferret that could annoy you as the owner is the tipping of food and water bowls that the ferrets do. This habit could also come as a shock to you. You might notice that the ferret just flips the food container or the water container. It has nothing to do with whether they like the food or not.

It is more like a signal to you that the ferret is bored. If the ferret is in the cage for too long, they can get lonely and bored. In their anger and boredom, they will just flip the container. You should make sure that the pet isn't caged for too long so as to keep him away from loneliness.

Apart from being bored, there is another reason that could be behind this action of the pet ferret. The animal loves to dig. Out of habit they dig and flip things around them. The ferret also likes to play in water, especially shallow water. And, they also like to play with everything around them. So, this could be their way of playing with the food and the water.

There are a few tricks that you can employ to keep this behaviour in check. You can't change the basic behaviour of the ferret, but you can work things around his behaviour. For example, you should use heavy base containers for food and water so that the ferret can't easily flip them.

You can also make use of extensions that keep the containers attached to the cage. This will also help to keep the containers in place. There is another trick that you can use to make the ferret eat the food rather than play with it.

Sprinkle some water over the food in the food container. This will attract the ferret. He will try to lick the water droplets and in the process will also eat some of the food. You should throw away this food if the ferret does not eat it because food mixed with water will get spoilt easily.

10). Digging in the litter box

As you might have understood by now, the ferret is very fond of digging up places. He might not leave the litter box. A ferret in a litter box is not something very hygienic. You will have to adopt some tips to stop your ferret from digging in the litter box.

If the box is very clean, it could be like an invitation to the ferret to come and dig. You can try keeping some faeces in the box, so that the ferret stays away from it. When the ferret approaches such a place, he will smell the stool. The smell will tell him that this place is to poop and not dig.

Another way to reduce the behaviour of digging anywhere is to get a sand box for the ferret. You will find it easily at a pet shop that houses ferret related products. This box will give the ferret a designated place to dig. This will reduce the impulse of the pet to dig any and every place.

11). Dooking

Dooking refers to the sounds that are made by the ferrets when they are very happy. An excited ferret will display its happiness by jumping around and also by making certain noises. Sometimes, these noises are very soft, and you will have to listen closely to even notice.

At other times, the ecstatic ferret will make very loud noises. The dance of joy is often accompanied by dooking. You should not be scared of these noises because they are perfectly normal. You should pay more attention to your pet's moods and behaviour to understand what kind of sounds it makes at different times.

12). Hissing

While dooking means that all is well with the pet, hissing is just the opposite. When the ferret is frightened or angry, it will make noises that sound like a hiss. This is known as hissing. Again, you have to closely observe your pet to understand his moods and the accompanying sounds.

When you see a sad or angry ferret, your first instinct as the parent would be to pick the pet up and calm him down. But, you are advised not to do so. When a pet ferret is angry, he needs some time to calm down. If you get over protective, he might just get irritated and might just bite you in frustration.

You should say some kind and soothing words to the pet. Your tone will make him realize that you care for him. After that just leave the area. The ferret needs some time. He will utilize this time and will cool down on his own.

Another point to be noted here is that an over excited ferret might also make a similar noise. When two or more ferrets are left to play with each other, they could emit a sound quite similar to hissing. It is important that you understand whether the pet is angry, scared or just having fun.

You can use a simple trick to know whether the pet is angry or not. You should pay attention to the ferret's body language. If the ferret is sticking to a corner, is making noises, has a bent or arched back and has his fur standing, then the ferret is definitely upset and needs time to calm down.

13). Nipping

Ferrets have a tendency to nip. You should know that this is absolutely normal for a ferret and that you can slowly train the ferret not to exhibit such behaviour. It is important that you understand the reason behind a ferret's nipping. You should not harm the pet when he nips. This could scare him and will make things worse for you.

More often than not, ferrets do so when they are in a playful mood. If your ferret wants you to play with him, he could just signal you to do so by nipping. Such behaviour is quite common in younger ferrets. So, don't be surprised when the young ferret nips really hard.

Nipping comes very naturally to the ferrets. In their natural environment, ferrets are known to nip each other. But, this does not harm them because of the quality of their skin. If you notice the skin of your pet, you will find it to be very thick. This thick skin is a cushion for the ferret.

Another reason behind a ferret's nipping is that the animal could be scared. When you bring the pet to your home for the first time, everything around him will be new. It is quite natural for the pet to get scared. This is the reason that nipping is very common in a new pet ferret.

14). Sneezing

Another trait that will catch your eye very quickly is the frequent sneezing of your ferret. A ferret is not very tall. His nose is very close to the ground, and on top of that the ferret has a habit of smelling everything. This can cause dust to enter their nose, hence the sneezing.

While sneezing is normal for a ferret, if you notice a running nose, then you should definitely consult a veterinarian. This could mean that the ferret has a cold, and if not treated on time, it can get very serious.

Ferrets have a tendency to catch bacteria and viruses very easily. So, you need to be on the lookout for various symptoms in them. If the cold of the ferret does not get better, he could develop mucus and coughing, which can be a very serious problem. You should never compromise on the health of the pet.

15). Scratching

You will also catch your ferret scratching itself multiple times. As a new owner, this could bug you. But, you should know that this is a very normal behaviour. The skin of the ferret is such that it can get itchy and the pet might have to scratch.

Though scratching is normal and nothing worrisome, if there are other symptoms that accompany it, then you need to see a vet. If the skin is red and shows abrasions, then this could be some kind of allergy. You need to check with the vet to understand the condition.

The animal also has a tendency to catch fleas. If you spot fleas on the animal, don't ignore it. You can make use of anti-flea products for the pet and should consult the veterinarian if the symptoms persist.

16). Scooting

There may be times when you will spot your ferret trying everything he can to move an object, even if it is too big for him. This will look particularly entertaining to many people as the ferret tries very hard, but in the end gets nowhere. Don't worry as this is just one of the plays of the ferret.

The pet will use his paws to hold on to the object and will get as close to the object as possible. He will then scoot backwards and will try to drag the object. He might even start running in a circle. Your pet is an endearing animal, and he will also be up to some mischief or the other.

17). Hiding objects

You might also find your ferret hiding certain stuff. The thing with the ferret is that they can get obsessed with certain things. For example, if there are ten toys in the ferret's cage, he might get too attached to a couple of them.

If he gets obsessed with something, he will not chew on it or harm it but will look for places to hide that thing. He might hide his favourite shirt or toy in a place where he feels that it is safe.

Another thing with these pets is that they will not like you or anybody getting close to that object. For example, if you find out the hiding place of the ferret's favourite toy and take it out from there, he will get very angry and also stressed. So, you should definitely avoid giving your pet this kind of stress.

Your job should be to make sure that they are hiding stuff that is safe for him and not too relevant for you. You can't allow the pet to hide a knife or cupboard keys. Always know what the pet is up to, so that he can't come up with something dangerous.

Make sure you don't let them get obsessed with such things. Break the obsession right in the beginning, so that the pet can look for a new thing to get obsessed and hide.

18). Wagging of the tail

Another trait that you can expect from your ferret is the wagging of the tail. This is a very lovable feature. Like most animals, the ferret will wag its tail out of excitement.

You can notice this behaviour in ferrets when they get along well with each other and when they are playing with each other. For example, the ferret will ferociously wag its tail when it is having a good play with another ferret the house.

Chapter 4. Setting the ferret's home

This chapter is an attempt to help you understand the importance of a shelter or a cage in an animal's life. You will be able to understand the basic concerns while building the cage for the ferret.

While it is important to have a cage, it is also important that the cage is of the right size. The advisable dimensions and specifications of the cage have also been listed. This will help you to build or a get a cage that is most suitable for your pet.

Like you need a home, an animal also needs a place and space that he can call his home. A home should make him happy and should be inviting for him. When the home does not provide the comfort and security that it should, it can lead to detrimental results.

There are many owners who might feel that there is no need to set up a cage because the pet can stay indoors. But, you need to remember that even if you are a hands-on parent of the pet, there will be times when the pet would be unsupervised.

There will be times when you will have to concentrate on some other work and the ferret would be alone. The cage is very handy at such times because you can do your work and can also be sure that your pet is safe and sound in the cage that you have built for him.

Also, during the night time, it is best for the pet and also for the family members that the pet sleeps in his cage. The pet will get used to the cage and your family members can also sleep without any tensions of your pet being loose in the house.

When you are setting up a cage for the ferret, you need to make sure that the cage is set up in a way that is inviting for the ferret. The ferret should not feel like a captive or a prisoner in the cage. If the ferret is not comfortable, he will begin to get stressed, which is something that you wouldn't want.

The cage should be built keeping in mind the basic nature of the ferret. You can't build a cage that is suitable for a tiger or a bird. You have a ferret and your cage should be built keeping in mind his natural behaviour, instincts, likes and dislikes. This is the best for you and also for the ferret.

You should understand that just because the ferret is a small animal does not mean that you can keep it anywhere. You need a proper cage for him. You should never keep him in a glass environment such as an aquarium. Such places don't allow the flow of air and can cause breathing issues in the ferret.

It is important to have the right temperature for the ferret. If the temperature is more than eighty degrees Fahrenheit or less than forty five degrees Fahrenheit, it is advised to keep the pet inside the house in controlled temperatures. Nothing is more important than the health and well-being of the pet.

When you will look for a cage for the ferret, you will realize that there are great options available for cages in ferret shops. You can buy a cage for as low as $75/£58.22 and also as high as $1000/£776.2. It clearly depends on your choice and your budget.

1). Building the right cage

If you are looking at the measurements of the cage then a cage of two feet height and three feet width is suitable for a ferret. This kind of a cage is suitable for up to three ferrets. The type of cage you have will directly affect the physical health and mental health of your pet ferret.

When you are building a cage for the ferret, you have to make sure that you have provisions for the most basic and important things, such as food and water. The ideal cage will be spacious enough. It will allow the animal to roam around freely and rest well when it wants to.

You can use two big containers for food and water. It is important that the pet has access to food and water at all times. You don't want to be busy somewhere else when your pet is stressed with the lack of water.

While you make sure that food and water is available to the pet, you also have to make sure that the containers are not movable. The ferret, owing to his natural tendency, might just kick the containers without realizing that the food and water in them was important for him.

To make things easier for you and the ferret, you can attach the food and water containers to the cage. You will easily get the tools to do so. When you attach the containers, the ferret can't move them. This will also help you to keep the cage clean and mess free.

The bedding that you choose for the ferret should be comfortable. It should not occupy the entire cage because your pet needs some space to roam around also. You can get the right bedding for your pet ferret from a pet store that sells ferret products.

The best buy for a cage is the one that can be cleaned easily. The cage should be comfortable and fun for the pet, but also easy to clean for you. You can go for a cage that has a bottom made of plastic and also coated wire. Such bottoms can be lifted for cleaning purposes.

But, make sure that the wire can't be chewed by the pet ferret. You can also go for the metal bottomed cage, but you need to be extra careful with these kinds of cages. You will have to make sure that such cages are not exposed to faeces and urine, otherwise they will rust. You can buy mats and rugs that can be thrown away after use to cover such bottoms.

A very important tip that you should always remember is that the cage bottom should not be covered with pine chips or cedar chips. These chips have certain oils that can cause damage to the liver of the animal and can also affect his respiratory system.

You might have seen ferrets being on display in some ferret shop in a cage with pine chips or a glass aquarium with the chips. But, if you wish to keep your ferrets healthy, both the glass boxes and pine chips should be avoided. The simplest cage that you can build for your pet is a cage with fixed food and water containers and mats over wire mesh flooring.

You should also keep the litter box and the food and water containers on opposite sides. The ferret would not want to defecate at a place where he eats food. You should also try to keep a hanging bottle of water along with the water bowl. This hanging bottle can be hung to the cage door.

You will have to take certain precautions with the litter box also. The pet will try to dig in a very clean litter box. To avoid this issue, you should be keeping paper litter in the box. You can easily get paper litter that is recyclable. Also, make sure that this is shredded or in pellet form.

If you get the clumping or the clay litter, these can get stuck in the nasal passage of the ferret when they try to smell the box. While getting the cage done, you should remember that certain amount of privacy is needed by the pet ferret. This is important so that the ferret can be healthy at a mental level.

If the ferret does not get what it wants, the pet will get stressed and might withdraw from you. There should be darkness in the cage when the ferret is fast asleep. You can even cover the top of the cage with a sheet or a blanket to make it really dark and secure for the pet ferret.

It is important that the bedding of the pet is soft and comfortable so that he can slide in and feel comfortable. But, make sure that you check the bedding every day to know whether the pet ferret has been chewing on its material. This can be dangerous so you need to replace such items.

If you wish to keep two or more ferrets in the house, then your cage requirements will automatically change. You can keep the ferrets in separate cages. If the ferrets get along, you can also keep them in a single cage.

You should make sure that the single cage is comfortable for both the animals. There should also be some additional space for the second ferret in the cage.

Though the ferrets are very small in size, they need to be comfortable in their shelter. It should be noted that you need additional space per animal in the cage. If you are planning to keep more than one ferret, then you should plan the additional space accordingly.

It is not advisable to keep too many ferrets in a single cage because it is not certain that all of them get along with each other. So before you plan the cage, make sure that you know how many animals would be sharing the space. This will help you to keep the right amount of additional space.

2). Accessories

Besides the basic stuff, such as food and water, it is also important to accessorize the cage well. This is important because the right accessories will help him to feel like he is at home. They will bring him closer to his natural habitat and natural tendencies.

When you are planning the furnishing and accessories of the shelter, you should make sure that you give the pet an environment that closely resembles his natural habitat. This will keep him happy and spirited. And when the pet is happy, then everything around is good.

When you are looking to place the bedding in the cage, you should remember that the ferret has a natural tendency to create and form burrows. He would want something that will help him to emulate the action of digging. There are several accessories available these days that will help you to keep your ferret happy.

When you bring a pet home, the pet will be scared of the new surroundings. You will have to make all the attempts that will help the pet to adjust in the new environment.

One of the safest ways to welcome a new pet is to provide him with a good shelter. The shelter should be as comfortable as possible. While you might save money on buying a cheap cage, you need to understand what is important for you.

The pet is more like a new member in the family, a new baby in the house. So when you buy the animal, you should make sure that you understand the needs of the animal at various stages of his life.

It is better to spend some extra money in the beginning than to see your pet being sad and lonely in the shelter. You should make sure that you understand this before you finalize on a cage for the pet.

A simple way to keep the ferret happy is to give him an old t-shirt or piece of cloth. The ferret will love it. He will act as if he is digging in the t-shirt. He will also try to fit in the t-shirt.

This will keep him busy and happy. This is one of his favourite plays. While the ferret enjoys the old cloth, you can bask in the happiness of your beloved pet.

If you go to a ferret toy shop, you will get many ideas for the accessories that you can keep in the cage of the pet. There are many types of bedding available these days that can help your ferret to rest and also have fun when he wants. For example, you can get bedding in the shape of a cave. This will be fun for the pet ferret.

You also have to make sure that the ferret is entertained even at times when you are not around. The ferret can get bored easily, which will make him a little aggressive. To keep him occupied, you can keep various kinds of toys in the ferret's cage.

The right kind of toys should be bought for the ferret. You will get many ideas when you visit a shop that keeps toys for ferrets. But, it is important that the toys are made of a good quality material. They should not be harmful for the pet. Your ferret will take them in his mouth, so they should be of a good quality.

It is better if the toys are washable. This will enable you to wash the ferret's toys every now and then when they are dirty. The harmful bacteria will also be removed from the toys when they are washed.

Also, make sure that the toys cannot be shredded by the ferret. If the pet is able to shred the toy, he will swallow the shreds. This is very harmful and will only invite more trouble for the pet. To avoid all these issues, buy the right kind of toys.

If you are planning to domesticate more than one ferret, you can consider buying another cage. This cage could be very simple and basic. The main purpose of this extra cage is to use it when one of the ferrets is sick. The cage will help you to isolate the sick ferret.

A vet will always advise you to isolate a sick pet. This is necessary so that the pet can recuperate nicely in absence of other pets. He would need some space to himself. What is also important is that he should not transmit the disease to the healthy pets.

3). Cleaning the cage

Like it is important to clean the house that you dwell in, it is extremely important to clean the cage of the pet. You will not necessarily enjoy this process, but still you have to do it. The pet can't clean the cage on its own, and if it is forced to stay in an unhygienic environment, he will fall sick.

There are certain tasks that you need to do daily, while several others need to be done once a week. If the bedding is soiled, it should be cleaned on a daily basis. Similarly, if the food and water containers look dirty, they should be cleaned and refilled. The litter box needs to be cleared every day.

Once a week, you should clean the entire cage. You should thoroughly clean it with a clean cloth. Remember that the ferret should not be in the cage when the cleaning procedure is going on. The litter box needs to be disinfected once a week. The toys of the ferret should be washed once every two weeks, if the toys are washable.

The litter box and the floor of the cage can be cleaned with the help of a mixture of bleach and water. The mixture should have 98 percent water and only 2 percent bleach. This daily and weekly cleaning procedure is important so that the surroundings of the ferret remain healthy. The bacteria in the dust and dirt can harm the ferret.

While you are busy cleaning the cage of the pet ferret, it is important that you check the cage thoroughly. If the ferret has littered in an area other than the litter box, then it should be cleared and disinfected properly. You should make it a point to do this check on a daily basis.

You can keep baby wipes handy to clean something immediately. It is important that the cage is free from all bacteria and viruses that are known to cause diseases in pet animals. You should keep some time designated for the cleaning of the cage.

If you are using bleach to clean the litter box, then you should make sure that there are no residues on the box. The ferret can try to lick on any residue that he may find on the box or the cage. Bleach can be very harmful and dangerous to the pet animal.

Another point that you need to understand is that you should not use very strong disinfectants. Such products can be very harmful if they ingested even in the smallest of quantities. You should always look for mild anti-bacterial soaps and detergents to clean the vessels and the floor.

A simple procedure that you can follow once every week to clean the cage thoroughly is to fill a bucket with clean water. Pour some anti-bacterial detergent that you wish to use. Form a nice lather in the bucket. This can be used to clean the toys and the containers. The remaining can be used to clean the floor nicely.

After you have cleaned the floor with the detergent, use plain water to wash off any sign of the detergent. This will ensure that the ferret does not ingest anything harmful.

It is also very important that you let the floor dry completely before you allow the ferret to come inside the cage. He could spoil the floor and could create a mess for you to clean again. He could even try to drink any residue that he finds on the floor. To avoid all these hassles, you should allow the floor to dry completely.

Chapter 5. Ferret proofing the house

When you have a ferret at home, you have to ensure that the pet is safe at all times. The ferret is so tiny that you might not know where he is most of time. This makes it very important that you understand the behaviour of your pet very well.

A ferret has a very curious personality. He will not think twice before charging into unknown territory. You might be busy with some work, and before you know your pet ferret might be walking into some real danger.

You should know that a ferret has a tendency to injure himself. If you don't pay attention, the damage could be very serious and irrevocable. A solution to keep your ferret safe is to ferret proof your home. This chapter will discuss the potential dangers to the ferrets and also some simple ways to ferret proof your house.

1). How to ferret proof the house?

This section will help you to understand the various ways to ferret proof your home. Make sure you ferret proof your home and keep your pet ferret away from potential dangers. There is no use to cry after the damage has been done. It is always better to take the necessary precautions in the very beginning.

To begin with, you should make sure that all liquid chemicals are far away from the ferret. If a chemical is in reach of the ferret, he might accidently spill it all over him. To make sure that nothing of the sort happens, you should make sure that all such supplies are kept in top cabinets where a ferret can't reach them.

You should also make sure that all kind of medicines, syrups and tablets are out of the reach of the ferret. These can be very harmful for the pet. You can also get childproof cabinets in your home to keep all such potentially dangerous stuff in those cabinets.

You would be surprised to know that your pet can climb toilets. Just imagine what can happen if the seat is not kept down. The ferret can slip inside and can get himself killed. To avert any such incident, make sure that the toilet seat is kept down.

You can also keep the toilet door closed to make sure that he does not enter the toilet. If there are any areas of the house that the ferret needs to keep away from, you have to keep them closed and blocked. If you don't do so, the pet can just enter the space when you are not around.

You should make sure that the ferret sleeps in his cage. This is for his safety and also for the good of the family members. You can also keep him in the cage when you can't supervise him and his actions.

You should make sure that your ferret plays with the right kind of toys. Cheap plastic materials that can have an adverse effect on the health of the ferret must be avoided. Similarly, toys that can be shredded or broken should also be avoided.

The ferret might accidentally swallow the small or shredded pieces. Make sure that the toys that you allow the pet to play with are of good quality. They should be safe for the ferret, and they should be impossible to swallow for the ferret.

Your ferret could actually shock you with the kind of things it can get hurt from. For example, the cardboard rolls of toilet paper can be very harmful for the ferret because he can get his head stuck in it.

You should make sure any such potentially dangerous things are out of the reach of the ferret. Keep the waste bin and waste stuff away from him because he might try to play with things that could be harmful for him. This might be very difficult for you in the beginning to look into areas and places that have hidden dangers for the pet.

But, you will definitely learn with time and experience. The furniture in the house should be ferret-friendly. You should make sure there are no sharp edges that could hurt the animal. Also, make sure that the ferret can't climb on the furniture.

If you have recliners in your house, keep them away from the pet. The pet could be severely injured by these reclining chairs. If somebody sits on them accidentally while the ferret is hiding in the spring, the reclining action and the spring could injure the pet. To be on the safer side, always check the chair or sofa that you are about to sit on. You don't want to sit on your ferret and injure him.

Your ferret could climb onto the washing machine and dish washer. So, make sure that these items always have a lid on. To be on the safer side, always check inside the washing machine and the dish washer before operating them.

The pet ferret should stay away from the plants of the house. You should also make sure that he stays away from Styrofoam products. All kinds of sponges should not be in the reach of the ferret. The ferret could bite into them and swallow them. This can be potentially very dangerous.

Rubber items can also be very dangerous if they are swallowed by the pet ferret. Imagine the kind of damage a rubber band can do if the ferret swallows it. You should know that the ferret will not know what he is not supposed to do. You will have to keep him away from danger.

Keep all rubber products locked up so that the pet can't reach them. This includes both soft rubber and foam rubber products. You should also make sure that soaps and detergents are kept away from the reach of the pet. These items can be very dangerous for the pet.

It is better to use the cabinets of bathrooms and rooms to keep things away from the pet ferret. You should make sure that the pet stays away from stacks of clothes. Keep the cupboards locked and keep the laundry area closed and locked. If the ferret gets inside a stack of clothes, you will have a very hard time finding him.

There could be so many things in your house that don't look dangerous, but could be very dangerous for your ferret. This is the reason that you might have to monitor the pet animal when he is not in his cage.

It is also a good idea to designate a spare room in the house for the ferret. This room should be open and spacious. It should have natural light and dim artificial lights. You can leave the ferret in the room and be sure that he is playing and having a good time. Of course, this will totally depend on whether you can spare a room in the house or not.

2). Areas of hidden dangers

While a ferret is a loveable and adorable pet, it can also be very destructive. As the owner, it is your responsibility to avoid any potential danger that the ferret can do. If a ferret is forced to stay in a cage for longer durations, he will get agitated and depressed.

You should allow the ferret to feel free in your home. But, to be able to do so, you will have to ferret proof your house. This means to create an environment where the ferret is happy and minimum damage is done to the things in the house.

Creating such an environment is essential to the ferret and also to the other family members. If you wish to make your home suitable for the ferret then you should be ready to let off certain things. You should not keep expensive carpets in areas where the pet will play because the ferret will spoil it. Instead, make use of old rags and carpets.

Don't keep breakable pieces in reach of the ferret. The ferret will approach it and might destroy it or hurt himself.

There are certain food items that are very dangerous for the ferret. You should make sure that the pet has no access to these items. Also, make sure that all chemicals and harmful substances are not reachable to him.

All the electrical equipment should be kept at a safe distance from the ferret. The sockets should be covered so that the pet animal is not harmed. You should not leave any food items on the table or shelves. He might eat it, without knowing whether it is good for him or bad for him.

There should be no wires on the floor of the house. The ferret can get entangled in these and can severely hurt himself. You should also keep all the doors and windows closed to not give him a chance to escape.

You should be very serious about ferret proofing your home. It is known that gastro intestinal disorders can cause a ferret to die very early in his life. Your ferret could just chew something dangerous and die.

If your pet ferret swallows something toxic for him, you might not even get a chance to take him to the veterinarian and save him. The digestive system of the animal is such that blockages can happen easily and they can be very dangerous. There are many ferrets that lose their lives because of such blockages.

This makes it very important to look for areas of hidden dangers and keep the pet safe. The ferret will try to chew anything it can. It will chew on rubber items, though such things are very harmful for him. It is you who needs to make sure that the pet does not chew on the wrong items.

Ferrets are fond of chewing on rubber items and foam and sponge based products. While they might like chewing on them, these materials when ingested will cause blockage of the digestive tract. You have to take measures to avoid such incidents in your home.

Your pet will love the stuffing of your sofa and couch. He will try to climb on the furniture and try to chew on the stuffing. He will also not miss an opportunity to chew on paper, cardboard and plastic items. These things are extremely dangerous for him if he swallows them.

If you don't keep an eye on him, your ferret will chew away all your expensive sofas. Additionally, they like digging and tunnelling. So, the animal will try to dig in the sofa material. You should be on the lookout of any tell-tale signs. If you see stuffing material on the floor, you should know what the ferret has been up to. He should be stopped as soon as possible.

Keep away your shoes in a cupboard or cabinet. Before you know, your ferret might start hiding in them and chewing on them. You should also check the bedding of the pet on a regular basis. If he has been chewing on it, then you should try a better and stronger fabric. These things are very harmful for the ferret.

Ferrets are also attracted to plants. They will merrily chew on the leaves of various plants. Many plants are known to be poisonous for the ferret, so they should not be encouraged to eat the leaves.

The best thing to do will obviously be to keep the plants in an area where the ferret can't reach them. This is a fool proof way to keep them away from the plants.

If you observe the pet closely, you will start understanding his likes and preferences. After you have kept an eye on him for a few days, you will start understanding his favourite spots in the house. You will know which areas he likes to dig and where he prefers to hide. These pieces of information can help you to ferret proof your home in a better way.

Your ferret will be very attracted to the smell of soaps and detergents. You need to be cautious when you are using store bought detergents and bleaching powder to clean surfaces and washrooms. A slight trace the bleaching agent can be harmful for the pet.

The ferret will go to a recently cleaned bath tub or toilet and lick the water off the surface. While this could be fun to watch for you, it could be dangerous if the tub and toilet still have traces of the detergent. To be on the safer side, you should always rinse the surfaces with excess water. This will make sure that the detergent has been washed off.

A ferret has a tendency to walk in to a situation and then not know what to do next. He might just climb on to the top of a closet, not knowing what to do. Ferrets will also crawl into any opening they see. For example, a ferret might get under the small opening of a fridge or refrigerator. This is very dangerous because the fan of the fridge can harm him.

Similarly, washing machines and dish washers are potential dangers to the animal. The best way to keep your pet animal out of danger is to know where he is and what he is up to. This will mean that you can help him if he has landed himself into some kind of danger.

You might also have to use barriers to make sure that the ferret can't reach certain spots and rooms in the house. But, a point that needs to be noted here is that normal pet barriers can't be used for a ferret. Even child proofing barriers would not be effective. This is because your ferret will happily climb these barriers. He might even get his head stuck in the barrier openings, inviting more trouble for himself and for you also.

You will have to make safe and secure barriers on your own. Or, you could get these barriers from a ferret shop. These barriers have a very strong base of plastic. Barriers made of Plexiglas will also serve the purpose right. If you wish to make the barrier at your home, then you can use Plexiglas or wood.

You can also use a good piece of cardboard as a barrier. For example, to keep the ferret away from the fridge or refrigerator, you can fix the cardboard in the opening. This will prevent the pet from entering the opening. Make sure you use good quality cardboard.

You can also take some measures to keep the ferret away from your furniture. This is important so that they don't try to create a tunnel by chewing on the fabric. You can fix some heavy material of cardboard at the bottom end of the furniture that you are trying to protect.

This can prevent the pet from digging on the material of the furniture. You can also keep such barriers in front of various rooms. This will make sure that the pet can't enter these rooms. These are simple ways to keep the pet safe and also your things safe.

It is important that you take appropriate steps to ferret proof the home. This will help you to set some limits and boundaries for the pet. These boundaries are for him.

There is no use in getting cautious after a serious and irrevocable damage has been done. It pays to take all the necessary precautions right from the very beginning.

It will take some time to understand the mannerisms of your pet. It is important to always supervise the pet. If you are unable to do so, you can ask a family member to do so for you. You can use the cage when there is no one around to supervise the pet ferret.

You can be secure knowing that you pet is safe and sound inside its cage. But, once you understand the pet better, you can take further steps to make sure that things remain safe in the house for the pet. A safe environment is good for everybody in the home.

Chapter 6. Diet requirements of the ferret

As the owner or as the prospective owner of a ferret, it should be your foremost concern to provide adequate and proper nutrition to the pet. If the pet animal in deficient in any nutrient, he will develop various deficiencies and acquire many diseases. When the nutrition is right, you can easily ward off many dangerous diseases.

The staple diet of an animal depends on its natural habitat and the food available around the habitat. Ferrets are known to feed on animal protein in their natural habitat. This is their staple diet.

When you domesticate an animal you can't keep it devoid of its natural and staple food. The digestive tract of the animal is also tuned to digest the staple food. You should always keep these points in mind.

Each animal species is different. Just because certain kinds of foods are good for your pet dog, it does not mean that they will be good for your pet ferret also. It is important to learn about all the foods that the ferrets are naturally inclined towards eating. You should always be looking at maintaining the good health of your pet.

It is important to learn about the foods that are good for your ferret. But, you should also understand that the foods that you feed your pet with could be lacking in certain nutrients. An animal in the wild is different from one in captivity. Availability of certain foods will also affect the diet of your pet.

Generally, the food given to captive pets is lacking in certain nutrients. It is not able to provide the pet with all the necessary nutrients. If such a case, you will have to give commercial pellets to your ferret. These pellets are known to compensate for the various nutritional deficiencies that the animal might have due to malnutrition.

You should always aim at providing wholesome nutrition to your pet. It is important to understand the pet's nutritional requirements and include all the nutrients in his daily meals. To meet his nutritional requirements, you might also have to give him certain supplements.

The supplements will help you to make up for the essential nutrients that are not found in his daily meals. Though these supplements are easily available, you should definitely consult a veterinarian before you give your ferret any kind of supplements.

It is very important that you serve only high quality food to your pet. If you are trying to save some money by buying cheaper low quality alternatives, then you are in a bad situation. A low quality food will affect the health of the ferret.

You can expect him to acquire deficiencies and diseases when he is not fed good quality food. The cure for this is taking the pet to the veterinarian. This in turn will only cost you more money. To avoid is endless loop, it is better to work on the basics.

1). Nutritional needs

The food that you give to your ferret should be capable of meeting its entire nutritional requirement. The food should appeal to the pet in terms of taste. He should want to eat it, and at the same time it should also provide the pet with all the necessary nutrients.

Meeting the nutritional requirements of the pet can be the biggest concern while looking to domesticate a pet ferret. You have to make sure that you understand the diet requirements of the ferret before you can decide to domesticate it.

A ferret that dwells freely in the wild is used to having a high protein diet. This protein basically comes from the carnivorous diet of the ferret. Ferrets love their meat. If you are planning to keep your ferret on a strict plant diet, then you are in for a big shock. These carnivorous animals need animal based protein in their diet. The food that you serve to the ferret should have at least 35 per cent of protein. It is better if the amount of protein is greater than 35 per cent.

The food of a ferret should also be rich in fats. The carnivorous diet of the ferret in the wild provides it with all the necessary fats. The fat content should be at least 20 per cent of the food. The protein and the fat in the diet of the ferret are suited to the kind of digestive system that it has.

The digestive tract of a ferret is quite small. This kind of digestive tract is suitable to digest good quality and easy to digest animal protein. It is not very suitable for food items such as corn, grain, and large quantities of vegetables and fruits. The main component of your ferret's main feed should be animal protein.

The ferret also needs the nutrient taurine. This is utilized for the cardiovascular system of the ferret and also for the eyes. The food that you serve your pet should have sufficient quantities of taurine. This means that you should keep the pet away from dog food because it is known to lack in this vital nutrient.

Also, the ferret needs more animal protein for maintaining its body functions. Dog food is known to be rich in plant protein, but not animal protein. Cecum is a part of the digestive system that helps the body to break down plant proteins. Ferrets don't have Cecum in their body.

2). Every day diet of the ferrets

If you are looking at buying food from the grocery store, then don't forget to check all the ingredients. If the main component is not meat, then you should leave the food item out of the ferret's diet. There will be many food items that will have cereal or corn or grain as the main ingredient. You should avoid such food items.

You should look for ferret foods in the pet shop to get the best food for your pet. Cat food or kitten food will not be the best buy for your pet. These food items will be unable to provide the right kind of nutrition to the ferret.

In case ferret food is not available in your area and you are only left with the choice of kitten and cat food, then you should consult a veterinarian for the right cat food. There are all sorts of cat foods available. You should buy the one that is the best for your ferret. Even in this case, you have to use fatty acid supplements along with the food to make up for the lack of fatty acids in cat and kitten food.

It is always a good idea to consult a vet in case of any doubts that you might be having regarding the diet and foods of the ferret. One of the best food items that take care of the protein and the fat content required by the ferret is 'Total Ferret'. It has over 35 per cent animal protein and over 20 per cent fat that is essential for the growth of the ferret.

The main component of the ferret's meal should be chicken. You can replace chicken with lamb if you wish to. If you are looking for ferret foods, then this should be the main ingredient of the food item. Don't forget to check the percentage of each nutrient in the food item.

Many a times, store bought foods are preserved by artificial means. You should not be buying such food products for your ferret. If the food item is preserved with the addition of Vitamin E in it, then this food is healthy for your ferret. It is also important that the food item is not adulterated with artificial colours.

As the parent of the pet ferret, you will be required to make smart choices in order to provide wholesome nutrition to your pet. Another thing that you need to make sure of while buying food for ferret is that the food should not have vegetables and fruits in dry form. This is extremely unhealthy for the pet.

Dried forms of vegetables and fruits are easily available; so many manufacturers might decide to use them. But, they can cause a serious health hazard to your pet ferret. They can get stuck in the digestive tract or intestine and can cause very harmful gastro intestinal issues. You should also avoid any food that has a high percentage of corn or grain or both.

If ferret food is not available, you might have to shift to cat or kitten food. But, this food would not be able to provide all the essential nutrients to your pet. There will more fibre than required and less quantities of animal protein. If you don't supplement the diet of the ferret, your ferret will suffer from some serious health issues.

Over the period of time, the skin of the pet will start to dull. This is one of the first signals that something is not right with the pet's health. The malnutrition will result in many other health issues. In severe cases, you can also expect kidney stones to be formed along with bladder stones. The reason for these stones is the excess fibre in the cat and kitten food.

If you have no other choice but to choose from cat and kitten food, then you should definitely go for kitten food. The reason behind this choice is that kitten food has a relatively higher percentage of protein when compared with equal quantities of cat food. Also, the fat content of kitten food is much higher than similar quantities of cat food.

It is also important to note that if a larger quantity of cat food is fed to the ferret for a longer duration of time, then the ferret can be diagnosed with insulinomas. This is the reason that you need to consult the vet before feeding the pet with any other food apart from ferret food.

In case you have been feeding the ferret with the wrong kind of foods, you need to switch as soon as possible. No matter how much damage has been done, you can still avert a lot of damage. So, never fail to make the right choices and right switches.

Your food choices will be limited by the area you live in. But, there is always a way to make sure that the optimal thing is being done for the ferret. There are many people who feed their pet ferret with cooked chicken along with standard kitten food. This makes sure that the ferret gets optimal amounts of protein and fat.

If you are also looking to feed the pet ferret with cooked chicken then you can look at serving the skin, heart, breast and liver. There are people who serve raw chicken to the ferret along with the kitten food. The problem with raw chicken is that it can introduce many disease causing parasites in the pet's intestine.

If you wish to save your ferret from these parasites then the raw chicken should be frozen. This will kill most of the disease causing parasites. Before you can serve the raw frozen chicken to the pet, it should be thawed properly. Another simple way to enhance the nutritional value of the pet's food is by allowing the ferret to chew and lick the softer ends of long chicken bones.

Letting the ferret feed on the bone marrow has a lot of advantages. It is a natural source of many needed nutrients. The animal will get its required quantity of Calcium from the bone marrow. Calcium is required for some essential body functions.

Another alternative that you can consider to supplement the main meals of the pet ferret is the easily available chicken baby food. You should make sure that this is only a supplement and not the regular meal. While you allow your pet to chew and the lick the bones, should make sure that the bones are of the right size.

If the bones are too small in size, the pet might just swallow the entire piece, leading to a detrimental blockage of the digestive tract. You should give bigger bones that can only be licked but not swallowed. In case you wish to feed the ferret with the big bones, make sure you boil these before serving them to the pet. This will make the bones softer for the pet.

It is always better if you can feed the ferret with high quality ferret food and also raw meat with it. A diet like this will take care of nutritional requirement of the pet. The pet should have access to water and also food at all times. You need not worry about the ferret over eating. It is known that these animals don't overeat. They eat as much as is required by their body, but they need frequent meals so it's better that they have easy access to food.

The pet would require almost 10 small meals in a day, owning to the fast metabolism of the ferret. The water can be kept in a container with a heavy base. Also, keep some water in a water bottle that can hang from the cage. You should change the contents of the containers twice a day. You might have to do it more than that if the pet has soiled the food or water while playing.

You may have to make changes to the ferret's diet as it grows older. It is known that older ferrets can have kidney issues. To lower the pressure on the kidneys, you should give the pet a diet low in animal protein. There are special diets for older ferrets, such as 'totally ferret for old ferrets'. You could also switch to a diet plan to maintain older cats. But, this should be done only if the ferret is above four years of age, and only after consulting the veterinarian.

3). Introducing new foods and switching foods

When you get a kit home, one of the biggest concerns that you will have is regarding its diet. It would take you time to understand the diet preferences of the new ferret. Ferrets form their preferences quite early in their lives. It is said that in the first six months, the ferret has his food preferences.

This means that the first six months is a great time for you to introduce different kinds of foods to the ferret. This will help him to have his preferences and will also make things easier for you. If your ferret likes three kinds of food items instead of one, it gets easier for you also.

In case a certain food item is not available, you know that you have other choices. If you don't introduce new foods to the ferret, he will turn out to be very fussy. And, you the parent will have a hard time keeping his taste buds happy.

If you want to introduce new foods or switch foods, you can't suddenly change his usual meal plan. This will put off the ferret. There are some simple tips and tricks that you should be following to make sure that the pet is eating well even when new foods are being introduced.

A simple way of introducing new food in the diet of the ferret is by starting out with a small amount of the food. Take a bowl and add the usual food of the pet in it. Now, take very small amount of the new food that you wish to feed your pet in the bowl. Mix the contents and serve the food to the ferret.

You should know that your ferret can out-smart you easily in this case. He will eat the usual food and might just leave the new food. This is because the new food item is alien to him. He needs time to get used to it. But, there is nothing to worry about, even if the pet leaves the new food in the beginning.

Just keep adding a very small amount of the food item in the usual food of the pet. Initially, the ferret will get used to the smell of the new food. This might take some time so be prepared to it. Once you see that the pet ferret has started easting the new food along with the usual old food, you can gradually increase the portion of the new food and decrease the portion of the old food.

Another trick to get the ferret accustomed to the smell of the new food item is by keeping the old and new food together. For example, you can store the two foods together in a container or in a zip lock. When you do this, the smells will get intermingled. So, even when you serve the pet with the usual food, you are serving him the smell of the new one.

The given process will take a few days, but you will have to have some patience. The idea is to help the ferret get used to the scent of a food before you can expect the pet to eat the food. Once he is okay with the scent, he will try out the food item on his own.

If you are looking for another trick to introduce new food items and switch between food items, then there is another one for you. Take some water in a bowl and add the two food items in the bowl. You can keep the quantity of the old food item a little more than the new one. Now, just stir this mixture and heat it for some time.

Don't heat the mixture of the food items for two long. You just need to heat it for a few seconds so that the smells blend with one another. If you still have your doubts whether your pet ferret will eat the food or not, then you can add 5-7 drops of 'Ferretone' in the bowl. This 'Ferretone', or kibble gravy, is loved by ferrets.

When you serve this food mix to the ferret, it is like a treat for him. He will waste no time in eating and licking the kibble gravy. In the process, he will also eat the contents of the bowl. This is a simple trick to make your pet eat new foods. This trick is also useful when the ferret is sick and is refusing to eat anything.

A point that should be noted here is that you should throw away the contents of the bowl that your pet does not eat. Don't keep it to serve him in the next meal. This is because the kibble gravy and water will cause the food items to spoil if they are kept for too long.

4). Treats

Treats are an essential part of a pet's meal plan. Treats are like small meal gifts that make the pet happy and delighted. The anticipation of getting a treat can also keep his behaviour in check.

You should work on giving your pet high quality treats. The treat should be tasty and also nutritious. The ferret should look forward to receiving a treat from you. This section will give you an idea of the kind of treats you can include in your ferret's meal plan.

It should be noted that just because your ferret seems to enjoy a treat, you can't give the food item to him all day long. You will have to keep a check on the amount of treats a ferret will get. This is important because treats are not food replacements. They are only small rewards.

It is also important that the pet ferret associates the treat with reward. He should know that he is being served the treat reward for a reason. You should also make sure that the treats are healthy for the pet.

If you keep serving him the wrong kinds of treats, it will only affect his health in the long run. This is the last thing that you would want as a parent of the pet.

This section will help you understand various kinds of treats that you can serve your pet. The best kind of treat for a ferret is a food item that has meat as its main component. Ferrets love their meat, and it is also healthy for them.

You should always look for treats that are healthy for the pet. The pet should enjoy eating them, but their nutrition should not be compromised. What is the fun of a treat if it is followed by multiple veterinarian visits? Your main aim should be to satisfy the pet's taste buds and also provide him some nutrition.

You should also make an attempt to understand what is in the treat that you prefer for the ferret. If you know the contents and their exact quantities, it will only get easier for you to make a well informed decision. Various pet shops will have ferret foods that will help you to make a decision regarding the pet's treats.

The treat should have the right mix of vitamins, fatty acids, minerals and proteins. This will make the treat healthy and wholesome. It is better if the treat has no sugar content. This is because the sugar will add no food value to the treat. Such healthy treats can be given to the pet ferret on a daily basis without any issues.

Be careful if you are planning to give your pet a bowl of fruit as a treat. If you think that a fruit or vegetable can serve as a treat for the pet, then you are absolutely wrong. The small intestines of the animal are not well equipped to digest such things very well. Though a small amount of these foods should be fine, a larger quantity will affect the health of the pet.

The pet can suffer from diarrhoea and other gastro intestinal problems because of consuming large amounts of vegetables and fruits. You will be shocked to know the problems that an undigested vegetable or fruit can cause in a ferret.

If there is a piece of undigested fruit or vegetable in the digestive tract of the animal, it can lead to obstructions and blockages. This will lead to many other digestive tract related complications.

The bowel movements of the pet can be restricted or completely stopped because of the undigested food. This can even pose a very serious threat to the life of the animal.

You should limit the consumption of fruits and vegetables to once or twice a week for the ferret. Even when you decide to give him these foods, remember to keep the quantity very low.

Also, you should make sure that you peel and mash the items before you serve them to the animal. This will allow him to digest the food well. There have been many reports of blockages in ferrets because of undigested carrots.

If you serve an uncooked carrot to the ferret, there is a very high chance that the ferret will be unable to digest it, leading to a urinary blockage.

The ferret needs to get its daily dose of meat, so you should include in some quantity in the treat also. Give 3-4 drops of Ferretone as a part of the treat. Your pet will love it.

You can also give him a small piece of Ferretvite. Remember to keep the quantity very low because this can increase the sugar content in the pets body and can also lead to toxicity.

You should keep in mind that you shouldn't serve something as a treat to the pet just because you like the food item. The food item should not harm the very sensitive digestive tract of the ferret.

As a rule, stay away from sodas, dairy products that are not for lactose intolerance, candy bars, chocolate pieces, caffeine, nuts and excessively salty and sugary foods.

These food items can cause some serious damage to the ferret. For example, if a nut gets stuck in the digestive tract, it can even kill the ferret. Excessive amounts of sugar can directly affect the work of the pancreas and the blood sugar level in the body.

Dairy products are known to cause gastro-intestinal issues in these animals. Also, a large amount of salt is unhealthy for the ferret. He can get really sick if you feed him with foods such as chips.

Look for ferret foods or kitten foods that can be served as treats. For example, Eukanuba Kitten food is a great example of a nutritious yet tasty treat for the ferret. Another example of a great treat is the Gerber chicken baby food.

Give your pet different kinds of ferret foods so that you have a lot of options to choose from. You can also buy small chew toys from the ferret store. The ferret will love this. You can also give him shreds of chicken or small pieces of boiled eggs as a treat.

5). Supplements

The diet of the ferret should be highly nutritious. If you make sure that the ferret is getting all its necessary nutrients from the food itself, you can avoid the use of supplements. At times, your ferret's diet might not be able to provide it with the right set of nutrients and vitamins. In such a case, it becomes necessary to introduce supplements in the diet of the ferret.

If the pet is not well and is recuperating from an injury or disease, the veterinarian might advise you to administer certain supplements to the pet. These supplements will help the pet to heal faster and get back on his feet sooner.

You should always consult a veterinarian before you administer any supplement to the ferret. He will be the best judge of which supplements the ferret requires and which ones he doesn't.

There are many vitamin supplements that are available in tasty treat forms for the ferret. While you can be sure that your pet is getting the right nutrients, the pet can enjoy the treat given to him.

You can also include supplements of fatty acids in the diet of the ferret. A few drops of this kind of supplement will enhance the taste and the nutritional value of the food item that is being served to the ferret.

While it can be necessary to supplement certain vitamins and nutrients to the pet, you should also be aware of the hazards of over feeding a certain nutrient. If there is an overdose of a certain vitamin in the body of the ferret, it can lead to vitamin toxicity.

Vitamin A toxicity is very common in ferrets. You should try to feed the ferret with an optimal amount of Vitamin A to avoid such a condition. You might even see that your pet is enjoying all the supplements, but this in no way means that you can give him an overdose. You should always do what is right for the ferret's health.

Another point that you should take care of is that you should not blindly follow the instructions and dosage that is printed on various supplements. The food that you feed the ferret will also have a supply of vitamins. The ferret will only require some extra dosage.

On a regular basis, you can look at giving the ferret treats with supplements, such as Ferretone and Ferretvite. These can be given on a daily basis, but the portions need to be controlled.

You can give 3-6 drops of Ferretone and pea-sized portion of Ferretvite. This is enough to supplement the daily requirements of the ferret.

Chapter 7. Health of the ferret

An unhealthy pet can be a nightmare for any owner. The last thing that you would want is to see your pet lying down in pain. Many disease causing parasites dwell in unhygienic places and food. If you take care of the hygiene and food of the ferret, there are many diseases that you can avert.

You should make sure that you do your best to prevent diseases by taking all the necessary precautions. If proper care is given to the food served to the pet, many diseases can be avoided.

You should always make sure that your pet ferret is always kept in a clean environment. A neat and clean environment will help you to keep off many common ailments and diseases.

You should make sure that the ferret has all his vaccines on time. Apart from this, you should take him for regular check-ups to the veterinarian. This is important so that even the smallest health issue can be tracked at an early stage.

At times, even after all the precautions that you take, the pet can get sick. It is always better to be well equipped so that you can help your pet. You should always consult a vet when you find any unusual traits and symptoms in the pet.

You should understand the various health related issues that your pet ferret can suffer from. This knowledge will help you to get the right treatment at the right time. It is also important that you understand how you can take care of a sick pet. This knowledge will help you to keep your calm and help the sick ferret.

1). Common health issues

Ferrets are prone to certain diseases such as adrenal diseases. You should know that the unique digestive system, fast metabolism and smaller size can cause the pet to get sick very easily. If proper care is not taken, you will find your pet getting sick very often.

This section will help you to understand the various diseases that a ferret can suffer from. The various symptoms and causes are also discussed in detail. This will help you to recognize a symptom, which could have otherwise gone unnoticed.

Though the section helps you to understand the various common health problems of the ferret, it should be understood that a vet should be consulted in case of any health related issue. A vet will physically examine the pet and suggest what is best for your pet ferret.

The various diseases that your ferret can suffer from are as follows:

Adrenal disease

Adrenal disease occurs when the adrenal gland in the body malfunctions. A ferret is prone to this disease. The main cause is growth of cells on a gland called the adrenal gland. This growth can present on the left or the right adrenal gland.

The growth on the gland can be both non-cancerous and cancerous. Though both the adrenal glands can possess such a growth, it is said that the left gland is more prone to this disease than the right one.

It is generally noticed that ferrets at the age of 3-4 years suffer from this disease. But, in many cases, ferrets below the age of three have also been known to suffer from the adrenal disease. The one symptom that you should never ignore is hair loss.

A pet suffering from this disease will shed lots and lots of hair. You should make sure that you consult the veterinarian as soon as you spot hair loss. Though the disease is very common in ferrets, this disease can be difficult to diagnose.

One of the major effects of adrenal disease is a change in the hormonal production of the ferret. This can often mislead many people. A drastic change in hormonal characteristics and production is often associated with Cushing's disease. But, this disease does not affect ferrets.

Another problem with the diagnosis of the adrenal disorder is that the blood reports will be normal for the ferret, indicating that all is fine with the animal. Even the X-ray reports will not show any issue.

There are some special blood tests that can be done for the ferrets to diagnose the adrenal disorder. You should make sure that your veterinarian has the facility to conduct these specialized blood tests.

It is important to understand that there are still many places and hospitals that don't have the facility to conduct these blood tests. In these cases, you will have to depend on the diagnosis of your veterinarian.

It is important that your veterinarian has the expertise to treat ferrets. This will help him to read the symptoms correctly.

The exact cause for this disease is still not known. According to one theory, the spaying and neutering of ferrets before they are sexually mature can lead to this condition.

There is another theory according to which the exposure to too much artificial light can lead to this health condition in ferrets. A ferret is not very comfortable in extremely hot or cold temperatures. This is the reason that most owners keep their pet ferret indoors.

Keeping the ferret inside a room with artificial sources of light disrupts the natural cycle of the pet to a great extent. The artificial lights can lead to this malfunctioning of the adrenal gland. Another reason behind this condition is high stress levels in ferrets.

You should try to keep your ferret away from too much exposure to artificial light. It is not possible to completely break the natural cycle of the pet, but there are a few precautions that you can take. The cage of the ferret should have no source of artificial light.

In the evening time or when the weather is pleasant, you should take the pet out in the open. Allow him to play in a safe area. This will expose him to some natural light. Also, when he is indoors, you can make sure that he is exposed to dull lights.

You should not keep the ferret in the dark and stress him out, just try to expose him to dull lights rather than exposing him to very bright artificial lights. The room should be exposed to the natural light cycles. It should not be artificially lit all the time.

If your kit is not spayed or neutered when you buy or adopt him, you should remember not to do it till the ferret is at least four to six months of age. A ferret will attain adulthood by six months, so it is better to wait till then.

You should also try to keep the stress levels of the ferret to a bare minimum. Don't put him in the cage at all times. Let him be in a ferret proof room of the house and also allow him to play in the open spaces of the house.

Symptoms:

Because the diagnosis of this disorder can be difficult, it makes it all the more important to read the symptoms well. You can look out for the following symptoms in the ferret to know that he is suffering from this particular disease:

- You will notice a sudden and drastic weight loss in the pet.

- There will be visible hair loss in the pet. If you observe the pattern of the hair loss, you will understand that it starts from the area of the tail. The hair loss moves up from there to the back of the animal.

- The skin of the ferret can get very flaky. You will also notice that the skin appears itchy and it will develop sores with time.

- Males and females can show some unique symptoms also. For example, you might notice that the male is very active during mating and otherwise he shows lethargy in his movements and actions.

- The female ferret will also show certain symptoms that can help you to diagnose this health condition. For example, the volva of the female will be swollen because of the growth on the adrenal glands.

Treatment:

In some cases, the animal is kept on a strict dose of certain medication. But, it should be noted here that this is not the permanent cure for the adrenal disease. This will definitely help to control the symptoms, but this is not a permanent cure.

If you wish to cure the pet fully then the affected part of the adrenal gland will have to be removed from the animal's body. Medication will have to be provided to help the other part of the gland to recover and function properly.

The disease and its treatment can have complications depending on the kind of tumour and also on the location of the tumour. If the outgrowth is cancerous, it automatically complicates the surgery. A metastasized tumour presents difficulties of removal, which complicates the process.

Also, if the affected adrenal gland is the left one, then the surgery is relatively simpler than when the right one is affected. The surgery on the right gland poses a danger to the major vena cava because the two are placed together.

There is a danger of rupturing the vein. If the entire right gland is not operated upon, then the symptoms can return.

Insulinomas

Low blood sugar conditions in ferrets leads to a condition called Insulinomas. When the ferret develops a condition in which he has an abnormal growth over his pancreas, that releases insulin, it is referred to as Insulinomas.

The release of this insulin in the blood stream leads to a condition of low blood pressure or hypoglycaemia in the animal.

It is also important to note that these growths could either be cancerous or non-cancerous in nature. If the condition is not treated, it can get very serious and can even lead to the animal's death. A ferret of the age of more than three is susceptible to this disease.

You should be on the lookout for the energy levels of the ferret. If he sleeps more than his usual sleeping hours and is lethargic, you should know that something is wrong. Talk to his veterinarian and get his blood test done.

The exact cause of this health condition is still not known. But, in most cases it has been noticed that this disease either accompanies or follows the adrenal disease.

Because this disease is related to the pancreas, this condition is definitely affected by the kind of simple carbohydrates the pet is fed.

The high amount of sugar in the diet of the ferret could also be a pre cursor to the disease. It is often advised that a ferret should be served with a protein rich diet.

You should limit the carbohydrates and sugar and increase the amount of proteins in the pet's diet. Even the treats that are served to the pet ferret should be healthy and not just sugar candies.

If you don't provide adequate amounts of protein in the diet of the animal, you will see him suffering from many ailments. There is no proof that a high protein rich can help you to avoid this condition, but it will definitely help in the ferret's growth and development.

Symptoms:

You can look out for the following symptoms in the ferret to know that he is suffering from this particular disease:

- You will notice the ferret to be very lazy and lethargic. It will appear that he has no energy to do anything.

- He will sleep a lot. If you try to wake him up, he can be unresponsive.

- Your pet will experience disorientation. He will not feel or seem coordinated in his actions or movements.

- The ferret will drool around and mouth and might also vomit occasionally. The pet will lose his appetite and will not show any interest in eating his food. He might also detest his favourite foods.

- Another symptom that can help you understand that your pet is suffering from Insulinomas is that he will experience seizures. You will notice sudden and jerky movements in the limbs of the pet. This could be accompanied by sudden passage of urine. He can also make sounds while he is sleeping.

Treatment:

One of the first things that you need to do when you see the pet suffering from a seizure is to apply Karo syrup all over his gums. This syrup will help the pet to come out of a seizure that is related to hypoglycaemia.

But, application of the Karo syrup is by no means a permanent fix to the ferret's problem. This is only a temporary relief to the poor pet.

Even after the temporary relief, you can expect another seizure very soon. This is because of the increased production of insulin in the ferret. You should visit the veterinarian as soon as possible to get the ferret tested. The vet will also give Prednisolone to make sure that the blood sugar is stabilized.

He might also suggest surgery to operate on the growth on the pancreas. But, a major problem is that Insulinomas can reoccur. There is no permanent cure for this health condition.

Another complication that can arise from Insulinomas is that it can spread from the pancreas to other organs. This can be very detrimental to the ferret's health.

In most cases, the vet advises surgery and continued medication for the pet. You would also have to ensure that you feed the pet with high protein and low carbohydrates in his meals. He should be fed frequent meals.

ECE

Epizootic catarrhal enteritis is a virus that is known to attack ferrets. The intestine of the animal is affected by the attack of this virus. It causes a disease that is called the Green slime disease. While you might think that a case of diarrhoea is not very serious, you need to watch out.

This virus can cause malnutrition and dehydration in the pet, which can take the shape of ulcers. A ferret that is old in age will be at the risk of ECE. But, this does not mean that a kit can't be attacked by ECE.

The symptoms in a kit are usually very subdued. Also, if the ferret is already suffering from an illness, he is likely to be going through this also.

Symptoms:

You can look out for the following symptoms in the ferret to know that he is suffering from this particular disease:

- One of the first symptoms of this disease is puking and vomiting in ferrets. Though there could be other reasons for the vomiting, you should not rule out ECE. The pet might discharge clear mucus at this time.

- The pet will soon get a severe case of diarrhoea. You can expect a green coloured stool during this time. The mucus causes this colour of discharge in the pet. Needless to say, the faeces will smell really bad.

- In case the condition of the ferret worsens, he will show symptoms of black faeces. The pet might grind his teeth from time to time and you will also notice red coloured spots in his mouth. These symptoms mean that the ferret is dehydrated and his condition is really bad. He might have developed ulcers. Take him to the vet for treatment without a delay.

Treatment:

Try to hydrate the ferret as much as possible. This will help him to feel better. You should look out for the symptoms in your ferret, and if the vomiting and diarrhoea episodes seem to increase with time, then you definitely need to see the veterinarian.

The vet will be able to guide you better in terms of medication for the pet. As stated earlier, there could be many other reasons for the prolonged vomiting and diarrhoea.

The vet will be a better judge of the symptoms and the condition of the pet. It might pay to keep the pet under his observation.

In the initial phase of an infection by ECE, the pet will feel nausea and will be restless. He might lose interest in eating and drinking water. But, it is more than important that the pet is well fed at this time. You will have to pamper him and make him eat and drink on time.

The vet might prescribe a special diet that will have to be followed. A point that needs to be noted here is that dehydration can often result in mouth ulcers. So, to avoid worsening of the condition of the pet, you should take care of his food and water.

No matter how much the pet resists, you have to make sure that he is being well fed when he is recuperating. If the ferret is dehydrated, you can give him Pedialyte with water. In severe cases, you can also give them Gatorade and water, but remember to dilute Gatorade with larger quantities of water.

The veterinarian will suggest the exact dosage of the mixes, which should be 15-20 millilitres of the mix in every four hours. If the pet ferret does not drink the mix on his own, you will have to syringe feed the animal. But, be careful while doing this so that the ferret does not develop any infection.

You should also make sure that the pet is eating a nutritious diet to allow his body to heal quickly. You can feed him Gerber chicken baby food, which is highly nutritious and light on the intestine.

You might have to hand feed him. You can use canned food or baby food that is prescribed by the veterinarian.

It is important that the food is easy on his digestive system. You can also feed the pet with Ferretvite along with the baby food. You will notice that the pet will get better with proper food and medication.

As the diarrhoea gets better, you might notice that the stool turns from the liquid to a grainy texture. This indicates that the pet is not getting adequate nutrients.

Talk to the vet and work out a diet plan for him where he gets adequate proteins. An important thing that you should know about the ECE is that even after your pet has recovered, he is still not free from the virus.

The pet himself will be safe from the condition after acquiring it once. It is known that the virus stays in the ferret's system for over six months. This is a reason that a new ferret is not encouraged to go near the older one for a very long time.

The new one could be carrying the virus, which he could easily transmit to your older ferret.

Gastrointestinal blockage

Gastrointestinal blockage is a very common problem in ferrets. It is even said that it is one of the main causes behind premature death in ferrets. This condition occurs when the ferret has a case of swollen intestinal tissue.

A ferret might accidentally swallow something dangerous for him, such as a foam or rubber piece. This will cause the blockage of the digestive tract in the ferret.

The best way to avoid such incidents is to always keep an eye on the pet. The ferret is a curious animal. He will always be running into some kind of trouble if you don't have an eye on him.

You should make sure that all dangerous items, such as rubber items and foam items are not in the reach of the pet. Keeping dangerous things out of the sight of the pet is probably the best way to avert all the tension that arises following a gastrointestinal blockage. Such a blockage ruptures the intestine tissue, making digestion of food very difficult for the pet.

Symptoms

You can look out for the following symptoms in the ferret to know that he is suffering from this particular disease:

- The pet will lose his appetite. You will find him avoiding even his favourite foods. He will not drink water, which could further lead to dehydration.

- You will notice a sudden and drastic weight loss in the pet.

- Another symptom of this disorder is vomiting. The pet will throw up from time to time.

- The pet would be seen struggling during his bowel movements. You should watch out for this symptom.

- You will notice the ferret to be very lazy and lethargic.

- The pet will suffer from diarrhoea.

Treatment:

If you find any of the above symptoms in a ferret, it is important that you waste no time and take the pet to the veterinarian. The vet will conduct an X-ray and ultrasound to confirm the blockage. Don't make the mistake of treating the pet at home.

Usually the symptoms start with vomiting. Severe dehydration follows the bouts of vomiting. If there is a blockage, the ferret would need surgery. It is important that you are mentally prepared for this.

Lymphoma

Your ferret is also at risk to another disease called Lymphosarcoma. This health disease is also called Lymphoma. It occurs because of the uncontrolled growth of the cells in the ferret's body.

Though this is very common in these animals, it can be difficult to detect, especially in the earlier stages.

Ferrets of age four and above are at the risk of Lymphoma. But, it is also known that a type of Lymphoma can also attack younger ferrets. You should never take any symptom lightly and should visit the vet when you observe changes in a ferret.

The vet will conduct tests on the blood sample of the pet to confirm this health condition.

Symptoms:

You can look out for the following symptoms in the ferret to know that he is suffering from this particular disease:

- You will notice a sudden and drastic weight loss in the pet.

- You will notice the ferret to be very lazy and lethargic. It will appear that he has no energy to do anything.

- The pet will suffer from diarrhoea. The lymph nodes of the pet will also be swollen.

- Another symptom that could accompany this disease is a cough. The pet will experience some difficulty in his breathing and will acquire a bad cough.

Treatment:

The treatment that is available for Lymphoma is chemotherapy. This in no way means that the pet will be disease free. The disease can reoccur in about seven months.

It is very difficult to save the pet after he has been diagnosed with this disease. Mostly, it gets detected in later stages, so the treatment becomes all the more difficult.

Because the symptoms of this disease are very general, it is suggested that you ask your veterinarian to conduct yearly tests for your pet.

This would help in detecting any issue in the very beginning, which makes it possible to treat it successfully.

2). Taking care of a sick pet ferret

If your ferret is sick, then it is very important that you take him to a qualified veterinarian. It is never advised to self-medicate. For example, if your ferret is suffering from a fever and you decide to give the animal a medicine that you take for fever, then you are in for a shock.

The medicines that work on human beings or other animals might not necessarily work on your ferret. You should never take this chance. Always consult the veterinarian before administering any medicine to the pet.

Along with the medication, you should also pamper the sick ferret. Ferrets love to be loved and pampered. You will see them recovering fast when you give them your attention and care.

While it will be a little difficult for you to take care of the ferret when he is sick, the experience can actually strengthen the bond that you share with the pet ferret.

The first sign that something is not right with the ferret is the body temperature of the animal. He should ideally be around 102 degree C. You should check this.

If you feel that the ferret is very warm to the touch, then you should know that the pet is not well.

Other symptoms that can help you to know that the pet is unwell include a lazy and lethargic pet. If you feel that the pet is not himself and has been acting very lazily, then this could be because he is unwell.

The pads of the feet of the ferret will also get warm in such a condition. This is another symptom that you should look out for.

When you see symptoms of a high fever in the pet, the first thing that you should do is make sure that the pet is drinking water. He can be given Pedialyte to help him recover. You should consult a veterinarian if the temperature does not come down in a few hours.

It is very important that you don't ignore the health condition of the ferret. Even if he has a slight fever, you should take it seriously because before you know the slight fever can take the shape of a life threatening disease. So, never hesitate to consult a vet in case of any doubts.

After you have consulted the veterinarian, you will have to spend a lot of time with your pet while he is recuperating from his illness. This can get very daunting for a new owner because you would not want to commit a mistake while taking care of your beloved pet.

You can take some simple precautions to make sure that your pet is healing better and faster. These precautions will ensure that the pet is getting all that is required for his healing process.

To begin with, you should always make sure that the pet is warm and comfortable. The pet will require something to curl into. He will also need his privacy at this time. Make sure that the ferret has a blanket or shirt that will allow him to do so.

Do not force him to do anything that he does not want to do. The pet needs some time and space. You should allow him to rest for as long as he wants. This will help him to heal in a better way.

If you have more than one ferret, then you need to keep the sick pet isolated. This is to give the ill pet time to get better and to avoid spreading the disease.

If there are things and toys that the pets share, you should wash these things nicely and keep them separately. You should wash the bedding and other washable accessories in the cage of the ferret. A pet recovering from surgery should be kept in a safe and closed environment so that he does not bruise himself.

A ferret can get dehydrated very easily. The pet might throw up when he is not well. This can easily lead to dehydration. The ferret will get disoriented if he is dehydrated for too long. You need to keep a check on the pet to ensure that he is not dehydrated.

You need to make sure that the pet is hydrated at all times. There is a simple way to find out whether the pet is dehydrated or not. You can pull the skin at the back of the neck of the ferret. If the skin does not fall back easily, then the pet is definitely dehydrated.

You should make sure that the pet is drinking enough water to get better. But, in some cases the pet might just refuse to drink any water. The vet might also suggest intravenous injections in severe cases.

You should make sure that that the pet does not consume cold water. This can cause severe diarrhoea in the pet. Water at room temperature is the best for the ferret.

If the ferret is extremely dehydrated, you can serve him Pedialyte with water to help him get better. In severe cases, you can also give them Gatorade and water, but Gatorade needs to be diluted with larger quantities of water because of its high sugar content.

These fluids will help the ferret to recover faster. Along with these mixes, your pet ferret should have access to simple water at all times.

The veterinarian will suggest the exact dosage of the mixes that your ferret needs depending on his condition. But, in general he should have 15-20 millilitres of the mix in every four hours.

A sick ferret will get dehydrated pretty quickly, so it is important to replenish his body with water and the lost salts. If you feel that the pet is not drinking enough water, you should consult your vet about the condition of the pet.

If the pet ferret does not drink the mix on his own, you will have to find a way to make him drink it. You can't force the pet, so the best way to replenish his body during a dehydration phase is to syringe feed him. If you take all the precautionary methods, this is not a difficult method.

To syringe feed the animal, take a clean syringe and fill it with the drink mix. Now, take this syringe to the side of the mouth of the pet. Slowly release a drop at a time in his mouth.

The ferret will not be so easy to feed, so you need to be patient. You need to be careful so that the ferret does not develop any infections.

The pet is not well, so you have to be prepared for making extra effort for him at this time. No matter how much the pet resists, you have to make sure that he is being well fed when he is recuperating. You should also make sure that the pet is eating a nutritious diet to allow his body to heal quickly.

A sick pet will also lose interest in eating his food. You might have to take out time and hand feed him. You can use canned food or baby food that is prescribed by the veterinarian.

It is important that the food is easy on his digestive system. The sick pet needs nutrition, but does not need the pressure of digesting heavy foods. If the food is difficult for the digestive system of the ferret to break, it will lead to more complications.

You can also feed the pet with Ferretvite along with the baby food. Ferretvite will help the pet to get the much needed taurine and vitamins in his diet. Make sure that the food is not very hot or cold. It should be warm and just right for the ferret.

Chapter 8. Grooming the ferret

When you decide to keep a ferret as a pet, you should understand that you will have to pay attention to the basic cleaning and grooming of the ferret. This is essential to keep the ferret clean and healthy. Not only will your ferret appear neat and clean, he will also be saved from many unwanted diseases.

When you are looking at grooming sessions for your ferret, you should pay special attention to the ferret's ears, nails, teeth and its bathing. This chapter will help you to understand the various dos and don'ts while grooming your pet ferret.

1). Ears

The ears of the ferret need to be cleaned regularly so that there is no wax deposited in the ears. There are many owners who might not consider ear cleaning an important part of ferret keeping, but in reality wax can lead to mite infestation and other infections.

In severe cases, the hearing of the pet can be compromised. It is important that you know of the early signs of mite infestation. The wax in the ears will have a light brown colour, while the wax with mites will be dark brown in colour.

It is important to see the veterinarian in case you have a doubt about mite infestation. Don't put any drops in the pet's ears without consulting the vet. In general, you should try to clean the ferret's ears once a week, or at least once in ten days.

You will require a cotton swab and an ear cleaning solution that is used for either ferrets or kittens. If there is somebody in the house who could help you, it will be easier to clean the ears.

If you are the only one doing this task, you should be calm and patient because ferrets don't like their ears being touched and cleaned. You can warm the cleaning solution before use.

Sit comfortably on the floor and hold the ferret gently by the loose skin behind the neck. Use your lap to give support to the ferret's legs. Take a cotton swab and apply some cleaning agent to it.

You should use the cotton swab with the cleaning agent to clean the parts of the ear that are easily visible to you. Don't go too deep because this can hurt the ferret. You should definitely not try to go further in the ear canal.

Repeat the process on both the ears. If you commit a small mistake from your side, it could cost the ferret his hearing. So, you need to make sure that whatever you do is gentle, yet done with firm hands.

The ferret might get uneasy and might try to get away from your grip. To make sure that the ferret is stable and not jerking, you can give him a treat. This will keep him occupied and will make your job easier.

2). Nails

It is important to cut the nails of the ferrets regularly. You should be looking at doing so at least once a month. If the nails of the ferret are not cut on a regular basis, there is a chance that the nails will get stuck somewhere. This will cause the nails to get uprooted.

You can imagine the pain your ferret will have to go through if the nails are uprooted. You will have to rush to the vet to help the ferret. Not only this, the long nails can also leave marks and scratches on your skin. So, make it a point to cut the nails of the pet regularly.

You should also make sure that you use the right equipment to cut the nails of the pet ferret. You should use good quality animal nail clippers. Along with that, you will need soap and styptic powder.

You can begin with giving a few drops of Ferretone to the ferret. This is to distract the animal so that he does not disturb you when you are busy clipping his nails.

If there is someone else in the house, you can ask them to hold the ferret. This will make your job easier. But, even if there is no one, you can do it on your own. Place the ferret in your lap in a way that he is comfortable and you have access to his nails.

You will notice a reddish vein on the nail. This is termed as the quick. You should cut the nail in a way that the quick is not touched. If you happen to cut it, it will hurt the pet and will also bleed.

In case you cut the nail in way that the quick starts bleeding, use the soap to clean it and then apply the powder. This will give relief to the pet. You should wait a few minutes for the ferret to feel better before starting the process of clipping the nails once again.

3). Teeth

When you are considering the overall hygiene and cleanliness of the ferret, you also have to take care of his teeth. You might have problems cleaning the pet ferret's teeth in the beginning, but he will get used it very quickly.

As a rule, you should try to clean the ferret's teeth once or twice a month. If you ignore his teeth, you will only invite unwanted problems for the ferret. You will notice tartar depositing on the teeth if they are not clean. This will automatically lead to decay of the teeth.

You should also know that many kidney issues in the ferret are also related to its bad oral health. So, it is better to be regular with the teeth cleaning procedure of the ferret.

It is also important that you take the pet to the vet if you see any tartar on the teeth. Even if all seems fine, it is advised to schedule dental check-ups for the ferret once or twice a year.

You can use toothpaste that is used for kittens and cats. You might also find toothpaste especially designed for the ferrets. Along with the toothpaste, you should use a soft brush that has been specially designed for kittens. A toothbrush with hard bristles might hurt the ferret's jaw, so you should avoid using it.

Your movements should be very soft. If you are too hard, you will hurt the ferret. Be very observant of the lather that comes out from the ferret's mouth. If you see a pink or red colour, you should immediately know that it is blood and that you are being too hard on the ferret's mouth.

Many owners complain that the ferret closes its mouth while the teeth are being cleaned. This makes it very difficult for the cleaning to take place. If your ferret does the same, then you can clean only one side of the mouth in one sitting. This means that you will have to be more frequent with the teeth cleaning sessions.

4). Bathing the ferret

Ferrets belong to the class of animals that are not extremely fond of bathing. Even within the ferrets, there are some ferrets that are okay with being in water and there are others who are hydrophobic. You will have to figure out whether your ferret is hydrophobic or not.

If your ferret is scared of water, you will have to try some tricks to get the ferret clean. Even if the ferret is hydrophobic, he needs to take a bath. This is something that you as the owner need to remember. The case of the hydrophobic ferret will be discussed later in this section.

It will be a difficult task for you to bathe your pet. But, this in no way means that it is okay for the ferrets to go without bathing. If the pet is not clean, he will attract fleas and other parasites. This only means extra work for you and veterinary visits for the ferret.

To avoid the ferret from getting sick, make sure that the ferret is bathed every now and then. The frequency would depend on the climate and the environment of the ferret. If it is too hot or if the surroundings are not too clean, it means that your pet should be given a bath more often.

Bathing is also important when the ferret is shedding its fur. The excess fur might stick on to the body. When you give the ferret a bath, the fur will just get washed off with water. This also means that the fur will not be shed all over the house.

Another point that you need to remember here is that while it is important to bathe the ferret once in a while, over-bathing is not recommended. This can also create many problems. The skin of the ferret will begin to lose many important essential oils if they are bathed frequently.

You will be surprised to learn this, but too much bathing can also increase the odour that the ferret might emit. To save yourself from these issues, try to keep things under control. As a rule, give your ferret a bath once in three or four weeks.

When you are looking to give a nice bath to your ferret, you should be looking at two things, a good quality mild shampoo and a few towels. It is very important that you choose the right shampoo for the ferret. If the shampoo is too hard or harsh, it will leave rashes on the ferret and might even cause serious damage to his skin.

You can buy a good quality cat shampoo or a baby shampoo for the ferret. These shampoos are very mild on the skin and have proven to be ideal for a ferret. You also need a few towels handy for the ferret. While one will be used to dry the water off, the others are required to cover the ground or floor.

If your ferret is suffering from flea infestation, then you will have to use a shampoo that can help the ferret to get rid of the fleas. You should consult the veterinary before you use a specialized flea shampoo. It is important not to take a chance on the health of the ferret.

If your ferret is not scared of water then it will relatively easier for you to bathe it. But, there are still a few precautions that you need to take. You should understand that how your ferret behaves under water will depend on its individual personality.

It is important that you make a few attempts to understand your pet's personality. Don't give up and understand his behaviour and mannerisms. This will only help you in your future dealings with the pet.

To begin with, make sure that the water you are using to bathe the pet is warm. Ferrets have a body temperature that is different from human beings. They should be bathed in warm water to keep them safe.

Take a tub and fill it half with warm water. Lift your ferret delicately in your hands. Make sure that your grip is firm. The ferret might surprise you when it touches water and might try to jump out of your hands. To avoid such a situation, place your hands on the stomach area and hold him firmly.

Place the ferret in the tub of warm water for a few seconds. Observe how he responds to water. If you see him enjoying, then your work becomes easier. You can also sprinkle water over the ferret. But, if the pet ferret is not enjoying then you need to be quick.

Take him out of the water, and put some shampoo on his back. You should form a good lather with your hands from the ears towards the tail region. Make sure that the pet does not escape when you are shampooing it. You need to have a firm grip on him.

You can also make use of the kitchen sink to give the ferret a bath. The sink will be deep and it will get difficult for the ferret to run away. If the ferret is hydrophobic, it is advised to use to two sinks or tubs. Fill both with water and use them alternately. Keep talking to your pet and make him feel that everything is fine.

You can also give him a treat at this point to divert his attention. The ferrets who are scared of water will give you a tough time. You have to be calm and quick. Keep the ferret in water only for a few seconds and keep alternating the tubs to divert the ferret and confuse him a bit.

Another way to bathe your naughty hydrophobic ferret is to sway him under running warm water. Turn the tap on and make sure the water is warm. It should not be cold or too hot. Once you are convinced that the temperature of the water is right for the pet, hold the pet and bring him under the water for a few seconds.

Before he starts to get fidgety, take him away from the water. Now apply some shampoo over the ferret. Keeping swaying him under the water till all the shampoo is washed off. It is very important that all the shampoo is washed off; else the ferret's skin will get infected and will show signs of rashes and abrasions.

While you are bathing the ferret, it is important that you protect his face. Water should not enter his eyes or ears. These are sensitive areas and water could cause some damage to them.

Keep him on the towels and use another towel to pat him dry. Make sure that he is absolutely dry before you let him go, otherwise dust and dirt will stick on his skin.

After the bath is finished, place the ferret in a big towel. You should place a few blankets or towels on the floor to keep it warm and tight for the ferret. The ferret will show too much energy at this time. He will try to escape you. You should be very gentle with the pet, otherwise you could harm him.

Chapter 9. Training the ferret

It is very important to train the animal to make him more suitable to a household. By nature, ferrets can be a little ferocious. You will have to train them to tame them.

You would definitely want the pet to be well trained. Though the ferret has been domesticated for years, you have to train him to make him more suitable to your home.

Training is required for all pets. Even a dog will need some kind of training. It is imperative to train a pet. This is a simple way to monitor their behaviour and to teach them what behaviour is acceptable and what isn't in your household.

If you wish to tame the pet, you will have to rely on some training skills to do so. These training skills will help you to have a control on the process of training. You will be able to monitor the progress of your ferret from time to time. This is important so that you know whether the pet is making any improvement or not.

Like training most other animals, ferret training will also require you to be patient. You will have to do a few trial and errors before you can be sure that your ferret is well trained. You should remember to have fun even during the training phase.

The training phase can be a great opportunity for you to learn more about your little pet. No matter how much you read about a ferret, your pet will have some individual properties that will separate him from the rest of the lot. This is a good time to learn about all these properties.

The more you learn about your pet, the stronger the bond you will form with him. You should remember to not take the training phase as a cumbersome thing. In fact, take it as an opportunity to form an everlasting bond with your pet. Your pet will also understand you better during this time.

While you have to be regular and stern during this phase, you should not be harsh and rude. Don't beat the ferret and terrorize him. You will only scare the pet and jeopardize your relationship with him.

If you have your doubts, it is better to read more about them and then make your decisions regarding the ferret's training phase. You have to display compassion and patience towards the pet if you wish to train him well, without scaring him for life.

As you would have understood by now, the ferret is a very curious and playful pet in nature. The curiosity of the ferret can be one of your challenges while looking to train him. But, at the same time the ferret is very intelligent, which should help you in your training.

When you are looking at training the ferret, you should be aiming for nip training and litter training above everything else. This training is important to help the ferret adjust into the household and also to make things easier for you and your family.

1).Nip training of the ferret

When you buy a new ferret, you might notice that the animal has a tendency to nip. This can be uncomfortable and worrisome for you as the owner. But, you should know that this is absolutely normal for a ferret and that you can slowly train the ferret not to exhibit such behaviour.

The first and foremost thing that you should remember is that you should not harm the pet when he nips. This could scare him and will make things worse for you. If you mishandle the pet and try to beat him, he might also try to bite you and harm you. Avoid going down this road and aim at training the ferret well.

It is important that you understand that reason behind a ferret's nipping. More often than not, ferrets do so when they are in a playful mood. If your ferret wants you to play with him, he could just signal you to do so by nipping. Such behaviour is quite common in younger ferrets. So, don't be surprised when the young ferret nips really hard.

Another reason behind a ferret's nipping is that the animal could be scared. When you bring the pet to your home for the first time, everything around him will be new. It is quite natural for the pet to get scared. This is the reason that nipping is very common in a new pet ferret.

When you know what you can expect from a new pet, it gets easier. Try to understand that he is still uncomfortable in the new surroundings and will require some time to get used to all that is new around him. Give him that space, time and also your understanding.

Nipping comes very naturally to the ferrets. In their natural environment, ferrets are known to nip each other. But, this does not harm them because of the quality of their skin. If you feel the skin of your pet, you will find it to be very thick. This thick skin is a cushion for the ferret.

The thick and tough skin of a ferret will protect against any nipping, but a human being does not have such a tough skin. The ferret might be happy and playful, but his nipping will hurt you, so it is important to train him against such behaviour.

As explained earlier, a ferret can exhibit such behaviour when they are scared. It should be noted that if the ferret has had a history of abuse, then you can expect him to nip more in fear than in a playful mood. If the ferret bites you very hard you can have a really bad wound. This makes it all the more important to train the pet.

There are many ferrets that are beaten up and abused. If you have rescued one such animal, then you will definitely find him trying to bite you out of fear and tension. But, don't worry because this is a passing phase. The love and warmth he will get at your place will help him to come out of his history of beatings and abuse.

If the pet is very young, he needs to be taught the behaviour that is expected out of him. He needs to learn to be sociable. He needs to learn that it is not okay to bite people. There are some tips and tricks that will help you to teach him all this.

Every time the pet tries to bite you, you should loudly say the word 'no'. Do it each time, till the ferret starts relating the word 'no' to something that he can't do. Don't beat him because this will only scare him. Just be stern with your words and also actions.

If think that the above trick is not very useful, then you can put the pet in his cage for some time. The pet will eventually understand that this behaviour will send him into the cage. The word 'no' and the act of putting him into the cage will make the pet more cautious of his behaviour.

It should be noted that it will take some time for the ferret to understand this. Until then just be patient and keep repeating these actions each time he tries to nip you. The ferret will look back on his memory eventually and relate the cage to something punishable.

Another trick to help the ferret understand that he can't nip and bite is to hold him and drag him away from you. Ferrets are used to dragging and pulling amongst themselves. They fight, nip and drag each other. The dominant one obviously wins. You need to establish the fact that you are the dominant one in the house.

When you are pulling the ferret away, you need to be very careful. You want to train the pet and not harm him. Use your thumb and the index finger to hold the skin at the back of the ferret's neck. This skin is loose and you will be able to hold it easily.

Look for the reactions of the ferret. He should not be in pain. The idea is to teach him to give up nipping and biting. When you hold him at the back of his neck, gently push him away from you. You might have to repeat this action several times before the ferret understands what is expected of him.

The ferret might also try to give you a good fight when you pull him away. Don't worry because this is something normal and quite natural of the ferret. The ferrets play and fight amongst themselves in the natural environment, so he might just try to defend and play with you.

There is another trick that can definitely help your training sessions with the ferret. You can apply something bitter on your toes and fingers, so that when the ferret nips you, he gets that bitter taste. When he gets to taste something bitter and terrible on you, he will eventually give up on nipping you.

It is important that the food item that you use is bitter but is not harmful for the ferret. You should know that there are some specially designed bitter foods for the ferrets. These food items are prepared keeping in mind the training of the pet ferrets.

You can easily buy these bitter food products online. The website everythingferret.com will help you with these products. You can also buy such products from the stores that have dog and cat related stuff. You can buy various bitter products, such as bitter apple and bitter lemon.

These products are extremely safe for the ferret, so you can use them without any doubts. The bitter taste will disgust the ferret. You just need to apply them or spray them to your toes and tips of the fingers. While you are working hard to train your pet ferret well, you should remember that you don't want to do anything that is not right for the pet in the long run.

For example, if you use too much of these bitter food products, the digestive system of the ferret can get upset. You just need to spray a little. This will be enough to get the job done and also not affect the ferret in a negative way. He just needs a little to get the bitter taste in his mouth.

After your pet has tasted the bitter product and is disgusted, you need to make it up to him. Wash off your hands and toes nicely and give the pet a treat. This is important so that the pet is not scared of you and your hands. This will also make him realize that nipping is not accepted, but eating from your hands is.

There are many treats that the ferret can lick. You can find these treats online. You can also treat your pet to these foods, so that he can affectionately lick from your hands. You should remember that the ferret will start ignoring and avoiding you if he only gets to taste bitter stuff from you. Be a teacher to the pet, but remember to be a friendly teacher.

Another point that you need to know while training your pet is that you need to monitor your actions too. You need to figure out whether nipping is a habit with the pet or has he suddenly started. If the pet has recently started nipping, then it could be something related to you.

It could be the smell of the new lotion that you started using or the smell of the new perfume that you started using. Ferrets can get attracted to certain kinds of fragrances and this can lead to nipping. You will have to figure out these things to train your pet well.

If the ferret loves a certain lotion that you wear, he will try to lick you and nip you just to get the flavour of that lotion. You can stop wearing the lotion to avoid such incidents. You also need to track what kinds of fragrances attract your pet more. This understanding will help you to train your pet well. You will know what can trigger your pet into nipping and biting.

The ferrets also have a strong tendency to nip the toes because sometimes when they play, they can only see your feet. This could be because of the fact they their eye sight is not that great. The feet could attract them. You might be surprised, but dirty and sticky feet will also attract the ferret. The ferret will try to bite and nip such feet.

A simple solution is to keep the feet clean and also wear socks in the house. When you wear socks, the ferret gets a little distracted. There are many owners of ferrets that have reported that wearing socks leads to a reduction of biting and nipping in the ferrets.

Another reason that could be behind your ferret's nipping and biting is that the ferret could be sick. You have to know your pet well to be able to detect sudden changes in his behaviour. If you see the pet being aggressive when you try to play with him, he could be sick.

You should thoroughly examine your pet for any injuries. If you spot an injury, you should take him to the vet. If he looks sick and tired, even then it is a good idea to take him to the vet. You should never postpone such things because this will drastically affect the pet's health.

While you are training your pet, you should remember that the ferret needs to feel comfortable and secure in your presence. You should spend quality time with him. Don't put him in the cage unnecessarily. If he left in the cage unattended all the time, he will become very aggressive. This will encourage his nipping and biting behaviour. Always remember that they can bite when they are scared and disappointed.

You should never neglect your pet. The ferret will learn slowly, but you have to be compassionate and kind towards the pet. Treat him when he exhibits good behaviour. This will encourage him further.

2). Litter training

A mother ferret will train the young kits to litter in one corner when they litter all around, but kits are weaned from their mothers very early. The mother does not get a chance to train the kits.

As the owner, you are also the care taker and the parent for the pet. You will have to teach him stuff that he needs to know when living in a family. Don't get upset when you see your ferret littering all around. You can train him not to do so.

To begin with, you should buy a few litter boxes. Keep these boxes in various areas of the house where the ferret is most likely to litter. You should cover the various corners where you have found the litter earlier. Also, install one box in the cage. Eventually, you want the ferret to litter in the cage itself.

It is believed that a ferret will generally litter in the first fifteen or twenty minutes of waking up. So, there is a chance that the ferret has already littered in the box in the cage. When you open the cage to take him out, check the box and wait till has used the box.

You should signal the ferret by pointing towards the litter box. The pet should slowly realize that he needs to use the box if he wants to get out of the cage. You should wait near the cage till he is all done.

Another point that you need to understand here is that ferrets are very smart. When the ferret understands that you will let him out of the cage once he uses the litter box, he might pretend to use it. You need to check the box and make sure that he has actually used it.

If you notice that the pet is not using the litter box installed in his cage, then you need to understand why. There is a chance that the litter box is uncomfortable for him. In such a case, you should look to buy a box with a front ledge that is low. This is good for your ferret.

You can even make one for your ferret. If you buy a cat litter box, you will notice that the front ledge is not too low. You could cut in into half to make it suitable for your ferret. The idea is to make it really comfortable for your pet ferret. A suitable litter box will have a back ledge that is high. This gives the right support to the pet.

The ferret will take its own good time to adjust with the environment. It is always difficult for a new pet to adjust. If you get him a new cage or if you make any changes in his surroundings, he will find it difficult to adjust. But, this problem is only time related and will get solved.

Every time the ferret litters outside the box, place his litter in the box that he should be using. You need to show the pet that he should be using the litter box. This could be difficult for you in the beginning, but the ferret will learn soon enough. You should place food and toys in areas and corners that you want to save.

When the ferret sees a toy or a food item in a corner, he will try to look for another corner to pass his stools. You can also place a mat underneath the litter box to save your carpet or home mats. Make sure that the mat that you use is water proof.

Observe your ferret's mannerisms when he is using the litter box. If he has a tendency to bite the mat underneath or stuff kept around, you should discourage this behaviour. To do so, you can use the bitter food sprays on the mats and other stuff. This will automatically discourage the pet from biting around when he is littering.

The litter box of the ferret should definitely be kept clean to maintain the overall hygiene and to prevent diseases. You should wash the box once a week. But, a point that needs to be noted here is that the box should not be too clean. A clean litter box that almost appears new could be appealing to you, but it is a turn off to the ferret.

The ferret will use its sense of smell to use the areas that he has used before. You should leave some paper litter in the box to encourage the pet to use the box again. This is a simple trick that you can use when you are trying to litter train your pet.

When you are buying a litter box, you should remember that the size of the box will depend on the size of your ferret. For example, a male pet ferret will need a bigger box due to his size as compared to the box that a female pet ferret will need.

If you are domesticating more than one ferret in your home, then this will also affect the littering process of the ferrets. This may come as a surprise to you, but the dominant pet could affect how the other pets use the litter boxes in the house.

You might notice that the habits of a dominant pet ferret are influencing the other pet ferrets. The dominant one will always try to boss around and make the others feel inferior.

Ferrets don't like to use the same litter box. The ferrets could also be competing for a litter box. These are the issues that you will have to find out. Observe which ferret is using which litter box and which one suddenly leaves a litter box.

You should make sure that each ferret has his own box, so that he not left to use the carpets and the floors. Even after you have trained your ferret to use the litter box, you have to be vigilant.

If you are observant, you might have to face issues. There could be instances when your pet ferret would suddenly give up the use of the litter box. Instead of getting angry with him, it is important that you probe into the reason for his sudden change in behaviour.

When the ferret is sick, he might give up the use of the litter box. The main reason behind this is that the pet might not have the strength in his hind legs to get on to the box. He could be suffering from a gastro intestinal or adrenal disease, which could make him weak and lethargic.

You should be cautious when you observe such changes in your pet ferret. Don't ignore his condition, or don't force him to use the litter box. You should not get angry with the pet because he is littering on the floor. It is not his fault if he is not well.

The best thing to do is such a situation is to take the pet to the vet. This will avoid the condition getting worse. He will look for the symptoms of various diseases and will help you to understand what is wrong with the pet.

3). Fun tricks with the ferrets

In the way you can train your ferret for various essential things, you can also train them to understand some of your tricks. For example, the animal can be trained to know that he is being called.

You can teach him to associate certain actions with certain words. There are many other fun tricks that will help you bond with your pet. These tricks are also very entertaining. Your family and friends will surely have a great time when you and your pet doing your fun tricks.

Like any other training, even these fun tricks will take some time. You will have to be patient with your pet if you want him to understand your commands well. There are some easy and fun tips that will make this process entertaining.

It can be very difficult if a ferret hides and refuses to budge. This situation can be a nightmare for any owner. What if the ferret got lost and couldn't find the way? To prepare yourself and your ferret better for such situations, you can try a simple and easy trick. Buy a few toys that make a sound for your ferret.

A ferret has a good sense of hearing so this will be helpful. You can buy these toys from the ferret shop or from the baby section. Make sure that the toys are made of a material that the ferret can't tear off and chew.

Use these toys one by one and make the respective noises. Pay attention to your ferret's reactions as to which sound he responds to the most. This will help you to take that one sound and proceed with further training of the ferret.

Keep this particular toy aside because it will help you a lot in the future. Use this toy and the sound to call your pet every now and then. The ferret will soon realize that when he hears the sound he should come to you.

This can be useful if you are unable to locate the pet ferret, which can happen quite often owing to the size of the ferret. You can just make the noise in different parts of the house. The ferret will hear the sound and come to you. This could help you to locate the pet.

Another simple trick that you can try is to make your pet do a little performance for you when you give him a special treat. Ferrets, like most other pets, love treats. Treats excite them and make them feel happy.

When he will see this favourite food item in your hand, he will get all excited. When you bring the treat near him, take the treat upwards. The ferret will try to reach the treat, and in this process will stand on the hind legs.

Next, lower the treat towards the ground. The movement of the treat in your hand will also inspire the movement of the ferret. Do it every time you give him the treat. Make sure that your actions are slow so that the ferret can follow them.

You can also say the word dance when you move the treat upwards and downwards. The ferret will slowly realize that to eat the treat he will have to stand on the hind legs and then sit back. You have to be patient to be able to teach such things to the pet ferret.

Make sure you do this only for fun and not to trouble your ferret. At the end of the trick, always give the treat to the ferret. Once you and your pet are good with the trick, you can flaunt and boast in front of your family and friends that you can make the pet dance.

This can be very entertaining for everybody who gets to watch this trick. Ferrets are anyways very entertaining and when you apply these simple tricks with them, they become more playful.

You can also teach your ferret to do a certain action for you. But, this will take a lot of time and patience from you. For example, you can teach the pet ferret to perform a roll over for you on the ground.

Lay the ferret on the ground, say the words roll over loud and clear and then give the pet a gentle roll over with your palm. Repeat this action many times.

The pet will start associating the word roll over with a roll over on the ground. When you say roll over, he will start rolling over on the ground on his own. This might take many weeks, even months of practice, so be prepared.

Also, if your pet seems uncomfortable with you rolling him over, you should just quit. Some ferrets might feel a little vulnerable with such an action. Understand your ferret's reactions and then take the next step.

Conclusion

Thank you again for purchasing this book!

I hope this book was able to help you in understanding the various ways to domesticate and care for ferrets.

Ferrets are adorable and lovable animals. These animals have been domesticated from many years. Even though they are loved as pets, they are not very common, and there are still many doubts regarding their domestication methods and techniques. There are many things that the prospective owners don't understand about the animal. They find themselves getting confused as to what should be done and what should be avoided.

A ferret is a small naughty animal that will keep you busy and entertained by all its unique antics and mischiefs. It is said that each animal is different from the other. Each one will have some traits that are unique to him. It is important to understand the traits that differentiate the ferret from other animals. You also have to be sure that you can provide for the animal. So, it is important to be acquainted with the dos and don'ts of keeping the ferret.

If you wish to raise a ferret as a pet, there are many things that you need to understand before you can domesticate the animal. You need to make sure that you are ready in terms of right preparation. There are certain unique characteristics of the animal that make him adorable, but these traits can also be very confusing for many people. You can't domesticate the animal with all the confusions in your head.

If you are still contemplating whether you want to domesticate the ferret or not, then it becomes all the more important for you to understand everything regarding the pet very well. You can only make a wise decision when you are acquainted will all these and more. When you are planning to domesticate a ferret as a pet, you should lay special emphasis on learning about its behaviour, habitat requirements, diet requirements and common health issues.

When you decide to domesticate an animal, it is important that you understand the animal and its species well. It is important to learn the basic nature and mannerisms of the animal. This book will help you to equip yourself with this knowledge. You will be able to appreciate the ferrets for what they are. You will also know what to expect from the animal. This will help you to decide whether the ferret is the right choice for you or not. If you already have a ferret, then this book will help you to strengthen your bond with your pet.

The ways and strategies discussed in the book are meant to help you get acquainted with everything that you need to know about ferrets. You will be able to understand the unique antics of the animal. This will help you to decide whether the ferret is suitable to be your pet. The book teaches you simple ways that will help you to understand your pet. This will allow you take care of your pet in a better way. You should be able to appreciate your pet and also care well for the animal with the help of the techniques discussed in this book.

Thank you and good luck!

References

http://www.nationalgeographic.com

www.ehow.co.uk

http://www.mnn.com

https://en.wikipedia.org

https://www.lovethatpet.com

http://www.ferret-world.com

https://www.thespruce.com

https://www.bluecross.org.uk

https://pethelpful.com

http://www.seniorlink.co.nz

http://www.drsfostersmith.com

www.bbc.co.uk

https://www.cuteness.com

http://www.arkive.org

http://www.vetstreet.com

https://www.hillsborovet.com

www.training.ntwc.org

www.wildlifehealth.org

http://animaldiversity.org

https://www.yourpetspace.info

http://healthypets.mercola.com

https://www.finecomb.com

https://a-z-animals.com

https://www.theguardian.com

http://www.businessinsider.com

http://www.kijiji.ca

https://www.gumtree.com

http://www.marshallpet.com

https://www.all-about-ferrets.com

Printed in Great Britain
by Amazon